The Forest Landscape
Restoration Handbook

The Earthscan Forest Library

Series Editor: Jeffrey A. Sayer

The Forest Certification Handbook 2nd edition
Ruth Nussbaum and Markku Simula

The Forest Landscape Restoration Handbook
Jennifer Rietbergen-McCracken, Stewart Maginnis and Alastair Sarre (eds)

Forest Quality: Assessing Forests at a Landscape Scale
Nigel Dudley, Rodolphe Schlaepfer, Jean-Paul Jeanrenaud,
William Jackson and Sue Stolton

Forests in Landscapes: Ecosystem Approaches to Sustainability
Jeffrey A. Sayer and Stewart Maginnis (eds)

Forests, People and Power: The Political Ecology of Reform in South Asia
Oliver Springate-Baginski and Piers Blaikie (eds)

Illegal Logging: Law Enforcement, Livelihood and the Timber Trade
Luca Tacconi (ed)

*Lessons from Forest Decentralization: Money, Justice and the Quest for Good
Governance in Asia-Pacific*
Carol J. Pierce Colfer, Ganga Ram Dahal and Doris Capistrano (eds)

*Logjam:
Deforestation and the Crisis of Global Governance*
David Humphreys

*Plantations, Privatization, Poverty and Power: Changing Ownership and
Management of State Forests*
Mike Garforth and James Mayers (eds)

Policy that Works for Forests and People: Real Prospects for Governance and Livelihoods
James Mayers and Stephen Bass

The Politics of Decentralization: Forests, Power and People
Carol J. Pierce Colfer and Doris Capistrano

The Sustainable Forestry Handbook 2nd edition
Sophie Higman, James Mayers, Stephen Bass, Neil Judd and Ruth Nussbaum

The Forest Landscape Restoration Handbook

*Edited by Jennifer Rietbergen-McCracken,
Stewart Maginnis and Alastair Sarre*

publishing for a sustainable future

London • Sterling, VA

First published in hardback by Earthscan in the UK and USA in 2007
Reprinted 2008
Paperback edition first published in 2008

Copyright © International Tropical Timber Organization, 2007

ISBN: 978-1-84407-584-3 (paperback)
ISBN: 978-1-84407-369-6 (hardback)

Typeset by JS Typesetting Ltd, Porthcawl, Mid Glamorgan
Printed and bound in the UK by TJ International, Padstow
Cover design by Susanne Harris

For a full list of publications please contact:

Earthscan
8–12 Camden High Street
London, NW1 0JH, UK
Tel: +44 (0)20 7387 8558
Fax: +44 (0)20 7387 8998
Email: earthinfo@earthscan.co.uk
Web: **www.earthscan.co.uk**

22883 Quicksilver Drive, Sterling, VA 20166-2012, USA

Earthscan publishes in association with the International Institute for Environment
and Development

A catalogue record for this book is available from the British Library

Library of Congress Cataloging-in-Publication Data has been applied for

The paper used for this book is FSC-certified and
totally chlorine-free. FSC (the Forest Stewardship
Council) is an international network to promote
responsible management of the world's forests.

Mixed Sources
Product group from well-managed
forests and other controlled sources
www.fsc.org Cert no. SGS-COC-2482
© 1996 Forest Stewardship Council

Contents

List of Boxes, Figures and Tables

Boxes

Figures

Tables

Preface

A whole range of services can be affected when ecosystems become degraded. Water quality can decline, carbon can be emitted into the atmosphere, biological diversity can be lost and the productivity of soils can decline. The deterioration of such services is felt most acutely at the local level, but it might also have implications regionally and globally.

Forest landscape restoration (FLR) provides a complementary framework to sustainable forest management and the ecosystem approach in landscapes where forest loss has caused a decline in the quality of ecosystem services. It doesn't aim to re-establish pristine forest, even if this were possible; rather, it aims to strengthen the resilience of landscapes and thereby keep future management options open. It also aims to support communities as they strive to increase and sustain the benefits they derive from the management of their land.

The term FLR is new, but most of its components are not. It combines adaptive management, participatory techniques and new (and not-so-new) technologies to create a flexible and creative approach to the use of trees in degraded landscapes. FLR also involves the use of a 'double filter', which implies that any FLR initiative should improve both the ecological functioning of a landscape and the well-being of the human communities that reside in that landscape. FLR takes a landscape-level view, which means that site-level restoration decisions should accommodate landscape-level objectives and take into account likely landscape-level impacts. It is important that it should be a collaborative process, involving a wide range of stakeholders to decide collectively on the best options for restoration.

This book has been written by a team of experts from a wide variety of institutions coordinated by ITTO and IUCN (see box overleaf). It explains the FLR concept and describes its main elements in chapters on adaptive management, landscape mosaics, landscape dynamics, stakeholder approaches, the identification of site-level options, hands-on site-level forest restoration and rehabilitation strategies, scenario modelling, and monitoring and evaluation. The result is by far the most comprehensive and easy to understand treatment of FLR yet written. It complements other work being carried out within the Global Partnership on Forest Landscape Restoration.

The **World Conservation Union (IUCN)** was one of the first conservation organizations to actively promote the idea of employing restoration as a conservation tool to complement the more established approaches of forest protection. If we are to address some of the major challenges facing natural resource management and conservation in the 21st century – for example, by contributing to poverty reduction and adapting to the impact of climate change – then we cannot afford to focus exclusively on large tracts of undisturbed forests. For this reason IUCN and World Wide Fund for Nature International (WWF) convened a meeting of restoration practitioners and experts in Spain during 2001, the principal outcome of which was a framework for considering restoration in terms of the broader landscape.

The **International Tropical Timber Organization (ITTO)** has been promoting forest restoration in the tropics for many years. As well as financing ten national-level workshops to further promote the guidelines and the FLR concept, ITTO was a major sponsor of the FLR Implementation Workshop in Pétropolis, Brazil, on 4–8 April 2005. In addition to its policy work, ITTO funds a wide range of forest restoration field projects in various countries in the tropics. In March 2003 ITTO joined the Global Partnership for Forest Landscape Restoration, of which IUCN was a founding member and currently acts as coordinator.

In 2002 IUCN and ITTO collaborated closely in publishing *Guidelines for the Restoration, Rehabilitation and Management of Degraded and Secondary Tropical Forests*, which also involved WWF, Food and Agriculture Organization (FAO) and Center for International Forestry Research (CIFOR). While these guidelines focus on conventional approaches to restoration, they incorporate most of the principles behind FLR, and their management and policy-level guidance is equally relevant to the implementation of an FLR approach. In 2003 and 2004 ITTO hosted a series of six regional workshops to make the guidelines more widely known in the tropics.

ITTO, IUCN and other partners will help deliver the messages contained in this book to forest restoration practitioners in tropical countries through a series of ten national-level workshops. It is our hope and expectation that this process will provide a major impetus to the implementation of FLR in the tropics and elsewhere.

Manoel Sobral Filho
Executive Director
ITTO

Ibrahim Thiaw
Acting Director General
IUCN

Acknowledgements

This publication presents the latest thinking on the emerging concept of forest landscape restoration (FLR). A joint production of ITTO and IUCN, it is the result of close collaboration between a number of institutions, including ITTO, IUCN, FAO, the Forestry Commission of Great Britain, WWF International, Intercooperation, CIFOR, the Japan Center for Area Studies and the University of Queensland, under the auspices of the Global Partnership on Forest Landscape Restoration (GPFLR). It has also drawn on the ideas and needs of tropical forest restoration practitioners. It builds on the ITTO *Guidelines for the Restoration, Management and Rehabilitation of Degraded and Secondary Tropical Forests*, which were published by ITTO in collaboration with FAO, Intercooperation, IUCN and WWF International in 2002.

The following people contributed to the development of *The Forest Landscape Restoration Handbook*: Edith Abruquah, Victor Kwame Agyeman, Stephen Aidoo, Isabelita Austria, Jürgen Blaser, Dominic Blay, Jill Bowling, Froylan Castañeda, James Gasana, Don Gilmour, Frits Hesselink, William Jackson, Wil de Jong, P. C. Kotwal, Trikurnianti Kusumanto, David Lamb, Eva Müller, Jean-Claude Nguinguiri, Duncan Ray, Ignacio Larco Roca, Hana Rubin, Cesar Sabogal, Sandeep Sengupta, Luis Tulio Salerno, Mario Javier Sanchez, Herman Savenije, Rudolphe Schlaepfer, Upik Rosalina Wasrin.

1

Introduction

Stewart Maginnis, Jennifer Rietbergen-McCracken and William Jackson

What is FLR?

The term 'forest landscape restoration' was first coined in 2001 by a group of forest restoration experts who met in Segovia, Spain, defining it as 'a process that aims to regain ecological integrity and enhance human well-being in deforested or degraded forest landscapes'.

The FLR concept is still being refined and redefined to accommodate new perspectives and ideas on what it entails and what sets it apart from more conventional approaches to putting trees back into the landscape. Indeed, the process of compiling *The Forest Landscape Restoration Handbook*, which involved discussions with a range of individuals and institutions, has itself brought increased clarity to the concept.

While the overall conceptual framework of FLR is new, virtually all the principles and techniques behind the approach have been around for some time and will already be familiar to many forestry practitioners. In essence, FLR is an approach to managing the dynamic and often complex interactions between the people, natural resources and land uses that comprise a landscape. It makes use of collaborative approaches to harmonize the many land-use decisions of stakeholders with the aims of restoring ecological integrity and enhancing the development of local communities and national economies. In many ways, it is an alternative to top-down, expert-driven land-use planning, providing a means to reflect societal choice through applying the principles of an ecosystem-management approach.

Thus, FLR differs from conventional restoration approaches in several ways:

- **It takes a landscape-level view.** This does not mean that every FLR initiative must be large-scale or expensive; rather that site-level restoration decisions need to accommodate landscape-level objectives and take into account likely landscape-level impacts.
- **It operates on the 'double filter' condition:** restoration efforts need to result in both improved ecological integrity and enhanced human well-being at the landscape level (the double filter is discussed in more detail later in this chapter).
- **It is a collaborative process** involving a wide range of stakeholder groups collectively deciding on the most technically appropriate and socio-economically acceptable options for restoration.
- **It does not necessarily aim to return forest landscapes to their original state,** but rather is a forward-looking approach that aims to strengthen the resilience of forest landscapes and keep future options open for optimizing the delivery of forest-related goods and services at the landscape level.
- **It can be applied not only to primary forests** but also to secondary forests, forest lands and even agricultural land.

The specific activities of any FLR initiative could include one or more of the following:

- rehabilitation and management of degraded primary forest;
- management of secondary forest;
- restoration of primary forest-related functions in degraded forest lands;
- promotion of natural regeneration on degraded lands and marginal agricultural sites;
- ecological restoration;
- plantations and planted forest; and
- agroforestry and other configurations of on-farm trees.

From policy to practice

The main aim of *The Forest Landscape Restoration Handbook* is to help forest restoration practitioners to understand FLR, appreciate its benefits and start to implement it. Thus, while the ITTO *Guidelines for the Restoration, Management and Rehabilitation of Degraded and Secondary Tropical Forests* (ITTO, 2002) are aimed primarily at policy-makers, this book targets field-level forest managers working in degraded forests and forest lands. These forest managers may include forest department staff, local communities or NGO staff involved in joint forest management, private sector timber company staff, or local government planning officers. FLR is still unknown to many of these groups, although they might already have adopted some of its principles in innovative forest restoration activities.[1]

One of the key messages in the book is that the technical knowledge is available to start FLR *now*, based on a wide range of proven restoration techniques. The limiting factors are most likely a lack of understanding of the landscape-level approach, the other land-use policies outside the forest sector that can have a major influence on landscape-level dynamics, and, in particular, the landscape-level impacts of site-level land uses. In addressing this last issue, the book highlights the double filter criterion of FLR, which states that the enhancement of human well-being and the restoration of ecological integrity cannot be traded off at the landscape level. This means that while specialization is inevitable and trade-offs unavoidable at the site level, the landscape-level sum of all site-level actions should attempt to balance the two objectives of enhanced human well-being and restored ecological integrity.[2]

This book has been compiled as a series of 'essential reading' chapters on the key principles and techniques of FLR and will serve as a bridge between the policy-level guidance provided by the ITTO guidelines and the context-specific field guides that it is hoped will be developed following the national-level FLR workshops to be held during 2005 and 2006. Though not a field guide, the book will still provide practical guidance on implementing FLR, including how to:

- use an adaptive management approach in planning and implementing an FLR initiative and support this approach through comprehensive monitoring and evaluation;
- understand and analyse the dynamics operating within a forest land-scape;
- work with multiple stakeholder groups and address different, sometimes conflicting, interests;
- construct FLR scenario models to help make explicit the choices and trade-offs inherent in FLR planning and facilitate collaborative learning with stakeholder groups on which technical options to pursue; and
- evaluate the technical options available at the site level and take into account the biophysical and socio-economic factors that will influence the likely success of an FLR initiative.

The book draws on numerous case studies in which FLR has been applied in practice (sometimes before the term FLR actually existed), and uses these to illustrate the main learning points on FLR. The book also provides references for further reading and more detailed guidance.

Notes

1   See, for example, the case study from the Shinyanga region of Tanzania in Chapter 2.
2   See Chapter 3 for more on the double filter of FLR.

Reference

ITTO (2002) *ITTO Guidelines for the Restoration, Management and Rehabilitation of Degraded and Secondary Tropical Forests*, ITTO Policy Development Series No 13, ITTO, Yokohama, Japan

2

What Is FLR and How Does It Differ from Current Approaches?

Stewart Maginnis and William Jackson

This chapter provides a brief overview of what FLR means in practice and what makes it fundamentally different from more conventional approaches to putting trees back into the landscape. The chapter highlights the following points:

- by itself, restoration through site-based interventions (such as afforestation schemes) is not capable of delivering the full range of forest goods and services that society and local communities require;
- the aim of FLR is not to recreate the past but rather to keep future options open, both in terms of human well-being and ecosystem functionality (including biodiversity conservation);
- land use and ecosystems change over time, so adaptability and, by extension, adaptive management lie at the heart of FLR; and
- delivering meaningful results at the landscape level will require more than just technically competent interventions; it will also require a good understanding of how land-use policies and people's livelihood needs influence the overall quality and availability of forest goods and services in the landscape.

To many foresters, the idea of a 'forest landscape' invokes an ideal image of continuous forest cover stretching uninterrupted towards the horizon, as illustrated in Figure 2.1. Under this scenario, the forest tends to be both well-managed and protected, delivering not only nationally and locally important products such as timber, rattan, fuelwood and rubber, but also maintaining important ecological services such as slope stabilization, hydrological regulation and carbon sequestration. In practice, however, the situation can be quite different.

Figure 2.1 *The 'classic' forest landscape: Lao PDR*

Source: Stuart Chape

Deforestation and forest degradation have altered many of the world's tropical forest landscapes to such a degree that, at the very most, only 42 per cent of remaining forest cover (or 18 per cent of original forest cover) in the tropics is still found in large, contiguous tracts. The forest estate of eight ITTO producer countries (and most ITTO consumer countries) now exists only as fragmented, mostly modified and sometimes degraded blocks. This means that at least 830 million hectares of tropical forest are confined to fragmented blocks, of which perhaps 500 million hectares are either degraded primary or secondary tropical forest and can be considered part of modified forest landscapes (Figures 2.2 and 2.3).

In addition to the large area of fragmented tropical forest, another 350 million hectares of former forest land can no longer be classified as forest because of the extent to which it has been degraded by fire, land clearance and destructive harvesting practices. Such areas, illustrated in Figure 2.4, often remain in a state of arrested succession because the conditions do not exist to support secondary forest regeneration or conversion to another productive land use. These areas lack nearly all forest-related attributes (structure, function,

Figure 2.2 *The 'secondary' forest landscape: Vietnam*

Source: Stewart Maginnis

productivity, composition) and constitute the greater part of degraded forest landscapes.

Finally, within modified forest landscapes in particular there exists an additional 400 million hectares of productive agricultural land which still retains a significant tree component.

Despite the fact that forest fragmentation, modification and degradation have shaped so much of the world's remaining tropical forests, many national forest strategies still tend to focus exclusively on how best to manage and protect intact areas of forest. And even when national forest programmes and strategies do recognize restoration as a priority, they tend to focus their restoration activities on the establishment of industrial roundwood plantations. Indeed, the fact that a natural forest no longer possesses all its original attributes has often been cited as a good enough reason to clear the area of its remaining vegetation and replace it with a planted forest. FLR builds on the growing realization that such strategies alone are insufficient to guarantee a healthy, productive and biologically rich forest estate in the longer term.

Figure 2.3 *The 'modified' forest landscape: Costa Rica*

Source: Alberto Salas

Figure 2.4 *The 'degraded' forest landscape: Papua New Guinea*

Source: David Lamb

Current responses to forest fragmentation and degradation

Plantation forestry very definitely has a place in FLR. However, afforestation alone cannot be expected to replace all the forest functions that have been lost or compromised through landscape-level deforestation, fragmentation and degradation. We therefore need to be realistic about what plantations are capable of delivering and recognize that space within the landscape needs to be created so that other complementary restoration strategies can be deployed. We also need to consider the 'multiple function' and 'dominant use' management paradigms not as mutually exclusive options, but rather as complementary options for use at different levels of forest management. 'Dominant use' is a perfectly legitimate approach to site-level activities, while the achievement of 'multiple functionality' should be the goal of landscape-level management. Thus a landscape configured so that it accommodates plantations, protected reserves, ecological corridors and stepping stones, regenerating secondary forests and agroforestry systems (or other agricultural systems that make use of on-farm trees) lays the foundation for multiple functionality.

What has been missing so far?

There is no lack of knowledge for implementing FLR; a wide range of experience has already been built up on how to restore some very difficult sites. What is more often lacking is an understanding of the overall landscape and the factors that determine whether different land uses (and land-use policies) are mutually reinforcing or in conflict. This landscape-level perspective is crucial if site-level decisions are going to contribute to an integrated restoration strategy. On the whole, it is forest management practitioners who take site-level decisions, and, while an enabling policy environment is necessary for successful FLR, such practitioners need not wait for the perfect policy before starting work. Indeed, progressive land-use policy is often based on experiences derived from innovative practice.

Box 2.1 Using a landscape perspective to enhance site-level management: Two case studies

Early attempts at large-scale reforestation of the Khao Kho district in central Thailand met with violent opposition from landless families, who often resorted to arson in order to prevent plantation establishment. The stand-off was resolved by looking at the broader issues within the landscape, incorporating local people into the project, reallocating about 500 hectares from reforestation to agriculture, and redefining the species mix and planting configuration to suit both local needs and technical challenges (Marghescu, 2001).

Oil-palm plantation managers along the Kinabatangan river in Sabah, Malaysia, observed that in some areas of their estate regular flooding prevented them from establishing an oil-palm crop. In collaboration with WWF and local communities, some of these managers encouraged secondary and planted forests to regenerate in affected areas, offering added protection to the rest of the estate while reducing fertilizer and pesticide run-off to the river, expanding species habitat and enhancing landscape connectivity for threatened species such as orang-utan and forest elephant, and optimizing the productivity of the flooded sites (WWF, 2002).

Taking a landscape-level perspective into account in site-level management results not only in potentially healthier landscapes, but also in improved stand-level management, as illustrated in the two case studies in Box 2.1. Both case studies highlight two key principles that are critical to building a landscape perspective into decision-making. These will be explored in greater detail in other chapters; for now it is only important to familiarize ourselves with what they are:

Meaningful public participation: reliable estimates indicate that there may be 500 million people living within modified and degraded forest landscapes in the humid tropics and dependent on a mixture of agricultural and forest resources to maintain their livelihoods. Practitioners need to realize that landscapes, especially modified or degraded ones, have many different stakeholder groups – each with their own particular needs and priorities. FLR seeks not only to take local people's needs into account but also to involve them actively in the decision-making process and subsequent implementation.

Balancing land-use trade-offs: it is common to hear about the need to pursue win–win solutions – that is, where two independent outcomes (such as biodiversity conservation and economic development) are maximized through a single intervention. In reality, however, win–win outcomes are extremely rare, particularly at the site level. There are often trade-offs involved between two sets of priorities and there is usually a need to develop compromise solutions. Without a landscape perspective, the same types of compromises tend to be repeated over and over again until key forest-related functions are lost from the landscape. The concern in the Khao Kho case study in Box 2.1 was that landless people's livelihood options would be compromised each time in favour of establishing new planted forest. In this case, restoration responded by ensuring that not all the forest area was planted and by modifying the species mix to ensure that local needs could be met.

In conclusion, conventional responses to fragmentation and degradation of forest resources can seldom be relied on to restore the full range of forest-related goods and services that society requires, since they rarely consider the broader landscape context or the livelihood needs of the people who live there. The rest of this chapter outlines how FLR can help practitioners respond to this challenge.

Defining FLR

There is nothing radically new about any of the individual elements of FLR. The approach draws heavily on a number of existing rural development, conservation and natural resource management principles and approaches that will be familiar to most readers. It has emerged as a concept over the last five years and is now being addressed by an established community of practice operating as the GPFLR[1]. The working definition used here is:

> *A process that aims to regain ecological integrity and enhance human well-being in deforested or degraded forest landscapes.*

Four key features of FLR are embodied in this definition:

1   ***FLR is a process***: implicit in the word 'process' are three key principles: (i) it is participatory; (ii) it is based on adaptive management and thus responsive to social, economic and environmental change; and (iii) it requires a clear and consistent evaluation and learning framework;

2 *FLR seeks to restore ecological integrity*: simply replacing one or two attributes of forest functionality across an entire landscape tends to be inequitable (as it caters to only a limited number of stakeholders' requirements) and unsustainable (as it is more difficult to respond proactively to environmental, social and economic change);

3 *FLR seeks to enhance human well-being*: the principle that the joint objectives of enhanced ecological integrity and human well-being cannot be traded off against each other at a landscape level is referred to as the 'double filter' of FLR; and

4 *implementation of FLR is at a landscape level*: this does not mean that FLR can only be applied on a large scale, but rather that site-level decisions need to be made within a landscape context. Some of the best examples of landscape-level restoration have been carried out with only relatively modest amounts of funding.

Technical components of FLR

This section provides an overview of the range of options that practitioners can consider when applying FLR. It is worth stressing again that the purpose of FLR is not to return forest landscapes to their original 'pristine' state, even if that were possible. Rather, it should be thought of as a forward-looking approach that can help strengthen the resilience of forest landscapes and keep future options open. It is important to understand that any individual application of this approach will be a flexible package of site-based techniques – from pure ecological restoration through blocks of plantations to planted on-farm trees – whose combined contribution will deliver significant landscape-level benefits. The site-level techniques can include:

- the rehabilitation and active management of degraded primary forest;
- the active management of secondary forest growth;
- the restoration of primary forest-related functions in degraded forest lands;
- the promotion of natural regeneration on degraded lands and marginal agricultural sites;
- ecological restoration;
- plantations and planted forests; and
- agroforestry and other configurations of on-farm trees.

Each of these techniques is outlined below.

The rehabilitation and active management of degraded primary forest: in degraded primary forests, the stand structure, composition, function and processes have been so adversely affected that satisfactory recovery of productivity and ecosystem integrity over the short to medium term will require active management interventions. Restoration in these cases would include removal of the causes of further disturbance and degradation, such as repeated annual fires, and promotion of stand recovery through targeted silvicultural treatments such as liberation thinning. Some of the most successful examples of degraded forest rehabilitation have been carried out by communities under collaborative forest management arrangements. Experience has shown that it is essential that communities are granted long-term rights to use both timber and non-timber products. Reneging on such arrangements once the forest has started to recover is not only unethical but also can be highly counterproductive.

The active management of secondary forest: secondary forest is woody vegetation that has re-established naturally on land that was previously cleared of most its original forest cover by shifting cultivation, settled agriculture, creating pasture lands or failed tree plantations (see Chapter 10 for a more comprehensive definition). These forest areas tend to be characterized by a relatively uniform composition of early successional species (i.e. pioneer and non-pioneer light-demanders), relatively even-aged stands and rapid initial tree growth. Many of these forests lend themselves to relatively productive monocyclic shelterwood systems over economically viable time-frames. This means that while they can rarely deliver all the attributes of an intact primary forest, they can, under certain conditions, provide a more ecologically attractive alternative to plantations. Because these forests are at an early successional stage they can respond well to silvicultural treatments such as liberation thinning. As in degraded primary forests, some of the most interesting management experiences have been those of local communities and small landholders.

Restoration of primary forest-related functions in degraded forest lands: unlike degraded primary forest, degraded forest land has been so severely damaged by worst-practice harvesting, poor management, repeated fire, grazing and other forms of disturbances and degradation that its vegetation cover can no longer be defined as forest. One example of this type of degraded land is the derived savannahs in the high-forest zone of west Africa that are dominated by *Imperata cylindricum*. Degraded forest lands are often highly dysfunctional in ecological terms, characterized as they are by low soil fertility and poor soil structure, soil erosion, the absence of fungal or root symbionts, and a lack of suitable micro-habitats for tree seed germination (due to the predominance of non-forest grasses and ferns and alien invasive species). In such situations, restoration activities are better focused on the recovery and maintenance of primary processes (hydrology, nutrient cycling, energy flows), rather than on attempting to replace the original forest structure or 'near-natural' species mixes immediately. As illustrated in Figures 2.5 and 2.6, hardy exotic species are sometimes the only option for site capture; these can then subsequently act as a nurse crop.

Promotion of natural regeneration on degraded forest lands and marginal agricultural sites: in some cases, degraded forest lands may still be capable of supporting natural regeneration. These lands tend to be of low productivity and can still be characterized as ecologically dysfunctional, though less so than the degraded forest lands described above. An example of this type of forest land is the low-productivity grazing pasture on laterite soils common in central America. Here, natural regeneration can be a viable

Figure 2.5 *Bamburi quarry, Kenya, was excavated down to 1m above the brackish water table*

Casuarina equisetifolia *was planted directly into small planting pits in the limestone with no other treatment (see foreground).*

Source: Stewart Maginnis

Figure 2.6 *Bamburi quarry, Kenya: The result after 20 years*

Soil structure is well developed, as is a native understorey. Casuarina *is being replaced by a native* Ficus *as the dominant canopy species and 19 IUCN red-list species have been recorded at the site.*

Source: Stewart Maginnis

proposition as long as the immediate drivers of degradation (such as recurrent fires or grazing pressure) are removed or carefully managed. The case study in Box 2.2 is a good example of what can be done under such circumstances. Two notes of caution, though. First, misdiagnosis of the drivers, processes and degree of degradation can result in major setbacks. For example, even if grazing pressure is removed from marginal pasture land, site recovery will be slow in the absence of desirable and viable seed sources. Second, one person's 'degraded' or 'marginal' land may be another's livelihood. Great care must be taken to avoid adversely affecting the poor and marginalized, whose principal source of income and sustenance may be these so-called 'degraded forest lands'.

Ecological restoration: given the scale of loss of some highly endangered forest types, many conservationists would like to see restoration efforts that aim to closely replicate the structure and floristic composition of the original forest cover, with intricate mixes of local tree species capable not only of site capture but also of attracting and sustaining local wildlife. Unfortunately, such intense ecological restoration at a large scale is a rare luxury as it is often prohibitively expensive, ecologically impractical and socially constrained. In some cases, even if these limitations could be addressed, this strictly defined ecological restoration will never be achieved because there is no reference ecosystem left from which to work. Nevertheless, ecological restoration can still be used judiciously to help create critical new habitat or connect existing fragmented habitats for endangered species and can be employed as one component of FLR. Indeed, major conservation benefits can be derived from combining ecological restoration with other restoration components, as illustrated in Box 2.2.

Box 2.2 Combining ecological restoration with other FLR components: A case study from Australia

In the tropical forests of north Queensland, planted forests have been used to add conservation value to ecological restoration across the landscape (Tucker, 2000; Goosem and Tucker, 1995). Faced with the challenge of creating new habitat on private farmland and restoring some semblance of landscape connectivity, the Queensland Parks and Wildlife Service has worked with landowners to restore critical biological corridors and 'stepping stones'. However, the fact that these corridors are no more than 100m wide and stretch over several kilometres of open countryside results in a large 'edge effect' that is highly unsuitable for species that require a 'deep' forest environment. This problem has been dealt with innovatively by planting commercial tree crops such as *Araucaria cunninghamii* adjacent to the restored corridor. The lessons generated in Queensland have broader application, not only for FLR on farmland but also as a conservation intervention in industrial plantations.

Plantations and planted forests: large-scale, industrial plantations and planted forests are not dealt with in great detail in this book, since ITTO guidelines already exist for best-practice plantation establishment and management (ITTO, 1993). However, it is important to reiterate that plantation forestry can be a key component of FLR. If properly designed and managed, even conventional monoculture plantations can make significant contributions to landscape-level biodiversity conservation and ecological integrity. Plantations can also play a catalytic role in the restoration and rehabilitation of degraded tropical lands by providing the necessary conditions for the establishment of native flora. In Porto Trombetas, Brazil, where a mixed plantation was established over an abandoned bauxite mine, at least 75 additional tree palm and shrub species were recruited naturally over the first ten-year period (Parrotta et al, 1997).

Table 2.1 sets out both the positive and negative impacts of plantations on FLR. Forward-looking managers of large-scale plantations will aim to increase the positive and reduce the negative. They will familiarize themselves with the concepts and ideas in this book and will feel encouraged to build a landscape perspective into their management decision-making processes. They can, for example, ensure that natural forest is maintained in riparian zones and utilize strips of natural forest to delineate compartments and working circles, thereby maximizing the contribution of new plantation schemes to landscape-level functionality.[2]

Agroforestry and other configurations of on-farm trees: on-farm trees are not only assets for farming systems but are also an important source of industrial roundwood and a means of enhancing ecosystem connectivity and maintaining landscape-level propagation capabilities. Indeed, some agroforestry systems are virtually indistinguishable from late successional secondary forest. Given recent FAO figures showing that agriculture continues to expand its land base in about 70 per cent of countries, it is highly likely that agroforestry systems will become an even more important component of FLR in the future.

Many of the challenges to making FLR work are not strictly technical but, indeed more often, social, legal and political in nature. For example, ambiguity over ownership rights for timber trees growing on private or communal agricultural land in Ghana during the 1980s and 1990s resulted in many farmers 'ring-barking' ecologically and economically valuable trees, making it almost impossible to persuade farmers to invest in tree-planting, even though this would have been beneficial from an agronomic point of view. Nevertheless, despite these kinds of problems, there is nearly always an opportunity for practitioners to take some decisions with a landscape perspective. Indeed, a fundamental characteristic of FLR is the use of a combination of technical approaches to solve problems, rather than relying on one particular type of intervention.

An FLR case study

This section looks at a case study that happened over a 20-year period and resulted in the restoration of over 3500km² of natural forest over a very large landscape. Although this experience happened before the concept of FLR was formally developed, it illustrates what FLR initiatives should be trying to attain. It also embodies the four key features of FLR outlined earlier in this chapter.

The Shinyanga region in Tanzania used to be covered with dense acacia and miombo woodland, but by 1985 much of the landscape had been transformed into semi-desert. Significant areas of forests had been cleared under colonial tsetse fly eradication schemes and some of the remaining areas were converted to cash crops such as cotton and rice in the 1970s. In 1975 many people were relocated under the government's 'villagization' programme, which meant that

Table 2.1 *Determinants of whether forest plantations
contribute to, or undermine, FLR*

	Positive	Negative
Environmental	• When the ecological functioning and productivity of degraded or biologically impoverished sites is improved • When conservation interventions are directed at the entire planted forest concession, not isolated sites • When the spatial design of planted forests emphasizes corridors and connectivity – especially between existing remnant habitats • When the species mix includes keystone food plants that accelerate wildlife colonization (especially ecological specialists) • When planted forests help maintain local genetic diversity	• When planted forests replace, simplify or isolate key species • When no provision is made to mitigate negative environmental off-site impacts such as run-off • When the planted species are, or create the conditions for the spread of, alien invasives • When planted forests or associated management systems significantly alter major ecological processes, e.g. natural fire regimes • When planted forests increase opportunities for ecological generalists at the expense of ecological specialists
Social	• When communities have a role in shaping the composition, location and configuration of planted forests • When people's rights to the forests and trees they plant are guaranteed and protected under law • When people's rights to places of cultural or spiritual significance are guaranteed	• When traditional access or use rights are disrupted or denied • When planted forests are established on disputed lands • When planted forest schemes reinforce rent-seeking behaviour by outsiders or local elites • When planted forests further disenfranchise marginalized sections of society
Economic	• When planted forests can contribute to the enhanced productivity of other land-use systems • When planted forests yield ancillary income-generating activities for local communities • When planted forest incentives can promote the delivery of multiple forest goods and services	• When planted forest incentives distort local and national markets • When planted forest incentives skew landscape-level trade-offs towards the supply of a very limited range of forest goods and services

they had to leave their homes, their farms and, most significantly, their *ngitili* – their enclosures of acacia-miombo woodland.

The Sukuma have long relied on *ngitili* to provide them with dry-season fodder for their cattle, firewood and other essential products. But by 1985 a mere 1000 hectares of *ngitili* remained across the entire region. Previous government land rehabilitation initiatives relied mostly on exotic species and had largely failed, so in 1985 government foresters started to consult with the local people as to what sort of strategy might be more likely to succeed. The response they received was almost unanimous – the restoration of the old system of *ngitili* should be a priority.

The first task of the new programme (HASHI, from the Swahili 'Hifadhi Ardhi Shinyanga') was to raise awareness about the importance of restoring forest resources within a degraded landscape context. Farmers and communities were helped to select the most promising sites for their *ngitili* and advised on how to manage them. Besides advising individual farmers, HASHI also worked closely with the *dagashida*, the traditional community assemblies that lay down and enforce customary by-laws. It wasn't long before the *ngitili* were transforming the lives of tens of thousands of people. In Mwendakulima village, for example, where animal fodder and forest product shortages were common, the villagers removed the grazing pressure from 105 hectares of severely degraded land in 1987 and the site was soon colonized through natural regeneration (see Figures 2.7 and 2.8). Income from *ngitili* is now regularly used throughout the Shinyanga region to support basic social services such as the construction of primary schools and the employment of local village health workers. In some villages there is anecdotal evidence that water supply has also improved because of *ngitili*.

The HASHI project recently surveyed 172 out of the 800 villages in Shinyanga region. They enumerated over 15,000 individual and communal *ngitili* covering around 70,000 hectares. This pattern of woodland restoration has also occurred in the 628 villages that were not surveyed, suggesting that it is highly likely that over 350,000 hectares of once-degraded forest land have been restored in a period of less than 20 years (Barrow et al, 2002).

Conclusion: what makes FLR different?

The concept of FLR is different from many other restoration-oriented technical responses for the following reasons:

- it focuses restoration decisions on how best to restore *forest functionality* (that is, the goods, services and processes that forests deliver), rather than on simply maximizing new forest cover. In other words, FLR is more than just tree-planting;
- it encourages the practitioner to take site-based decisions within a *landscape context*, ensuring, at the very least, that such decisions do not reduce the quality or quantity of forest-related functions at a landscape level and,

Figure 2.7 *The Shinyanga Region* ngitili, *mid-1980s*

In the mid-1980s it was estimated that the Shinyanga Region had only 1000 hectares of
ngitili. *At that time the landscape was typically barren and degraded, with few if any forest
resources.*

Source: Stewart Maginnis

    ideally, that the decisions contribute towards improving landscape-level
functionality;
- it requires that **local needs** are addressed and balanced alongside national-
level priorities and requirements for reforestation, thus making **local
stakeholder involvement** in planning and management decisions an
essential component;
- while promoting the need for site-level specialization, it strongly discourages
actions that would result in human well-being being traded off against
ecological integrity at the landscape level, or vice versa. Such trade-offs
are unsustainable and tend to be counterproductive in the medium to long
term;
- it recognizes that neither the solutions to complex land-use problems nor
the outcomes of a particular course of action can be predicted accurately,

Figure 2.8 *The 17-year-old Mwendakulima* ngitili

These villagers used an FLR approach to restore 105 hectares of productive woodland, mainly by excluding cattle from the area and silvicultural treatments.

Source: Stewart Maginnis

especially as ecosystems and land-use patterns change over time. FLR is therefore built on adaptive management and requires that necessary provision be made for monitoring and learning;

- given the complex challenge of restoration, FLR will normally require a package of tools, such as those discussed earlier in this chapter. Single-solution approaches will seldom provide the practitioner with sufficient flexibility; and

- finally, over the long term, FLR cannot be driven solely by good technical interventions alone but will require supportive local and national policy frameworks. In many situations it is likely that policy change will follow on from good innovative practice. Therefore, practitioners need to familiarize themselves with how other land-use policies impact on the restoration and management of forests.

Notes

1   The GPFLR is a network of governments, organizations, communities and individuals who recognize the importance of FLR and want to be part of a coordinated global effort. The facilitating partners are WWF, IUCN and the Forestry Commission of Great Britain. For more information, visit www.unep-wcmc.org/forest/restoration/globalpartnership.
2   More information on the role of planted forest in FLR is provided in Maginnis and Jackson (2002).

References and further reading

Barrow, E., Timmer, D., White, S. and Maginnis, S. (2002) *Forest Landscape Restoration: Building Assets for People and Nature – Experience from East Africa*, IUCN, Cambridge, UK

Ecott, T. (2002) *Forest Landscape Restoration: Working Examples from 5 Ecoregions*, WWF International/Doveton Press, Bristol, UK

Goosem, S. & Tucker, N. (1995) *Repairing the Rainforest: Theory and Practice of Rainforest Re-establishment in North Queensland's Wet Tropics*, Cassowary Publications, Wet Tropics Management Authority, Cairns, Australia

ITTO (1993) *ITTO Guidelines for the Establishment and Sustainable Management of Planted Tropical Forests*, ITTO Policy Development Series No 4, ITTO, Yokohama, Japan

Maginnis, S. & Jackson, W. (2002) 'Restoring forest landscapes', *ITTO Tropical Forest Update*, vol 12, no 4, pp9–11

Marghescu, T. (2001) 'Restoration of degraded forest land in Thailand: The case of Khao Ko', *Unasylva*, vol 52, no 207, FAO, Rome, Italy

Parrotta, J., Knowles, O. and Wunderle, J. (1997) 'Development of floristic diversity in 10-year-old restoration forests on bauxite-mined site in Amazonia', *Forest Ecology and Management*, vol 99, pp21–42

Tucker, N. (2000) 'Wildlife colonisation on restored tropical lands: What can it do, how can we hasten it and what can we expect?', in S. Elliott, J. Kerby, D. Blakesley, K. Hardwick, K. Woods and V. Anusarnsunthorn (eds) *Forest Restoration for Wildlife Conservation*, ITTO, Yokohama, Japan, and Forest Restoration Unit, Chiang Mai University, Thailand

WRI website (for forest-cover maps): http://forests.wri.org

3

Building Support for FLR

William Jackson and Stewart Maginnis

Successful FLR requires supportive local and national policy frameworks and a strong constituency of local-level support for the restoration activities. FLR therefore needs to include the identification of stakeholders and their forest-related interests and a consensus-building process on the range of possible restoration options. To achieve this, practitioners need to openly communicate and engage with a range of stakeholder groups. This chapter focuses on how FLR can help forest managers respond to local livelihood needs, build trust with local communities and demonstrate the importance of forests to decision-makers within a broader land-use and economic development context. The chapter also looks at how practitioners can make a strong case for the benefits of FLR based on its contribution to poverty reduction, economic growth, environmental security and biodiversity conservation.

It may seem odd that this topic is dealt with so early in the book. It is treated here as an up-front issue because without strong support from policy-makers, local communities, the private sector and other stakeholder groups, landscape-level restoration will not succeed.

Why foresters need to make forests more relevant

Degraded and secondary forests are a common feature in many tropical landscapes and are the main providers of forest-related goods and services. Their location near human settlements and the fact that they are often considered dysfunctional and unproductive means that these forests tend to be under greater threat than more isolated blocks of intact primary forest. Yet the complete loss of these forests would represent a further impoverishment of already degraded and modified landscapes and would extinguish any possibility of improving landscape-level functionality in the immediate future. From an economic perspective, the loss of these forests would result in the decline of an area's timber commodity base, with ensuing loss of jobs and livelihoods, while from a conservation perspective it would mean a local, possibly permanent, loss of forest biodiversity. If degraded and secondary forest resources are to be safeguarded and restored, decision-makers, local communities and the private sector all need to understand why they should support an activity that will take at least ten years, often much longer, to yield demonstrable dividends.

Given that FLR is underpinned by multi-stakeholder consultations and dialogues, it must begin by mobilizing support, particularly from local stakeholders. It is no accident that the first objective of the ITTO guidelines is to 'attain commitment to the management and restoration of degraded and secondary forests' (ITTO, 2002). Furthermore, convincing policy-makers of the value of FLR is important not only for the success of restoration initiatives but also for continued support for forestry activities in general. This is all the more important given the current context of declining funds for forestry. Indeed, unless foresters can start to convince their own governments of the real value of forests and the need to restore degraded forest landscapes, then it is likely that forest department budgets will decline significantly. Thus, although communication and advocacy skills are not usually taught in forestry courses, they are essential tools for any forestry practitioner.

The double filter of FLR

One of the key points highlighted in Chapter 2 was that FLR recognizes the need to enhance human well-being and restore long-term ecological integrity at the landscape level. This principle has been referred to as the 'double filter' of FLR. In looking to build support for FLR, this basic principle provides an excellent starting point as it incorporates the idea of pragmatic flexibility and provides clear guidance for the strategic direction of any restoration activity.

A common problem with any form of natural resource management is how to find the right balance between exploiting the resource for economic benefit and conserving the same resource for environmental protection, biodiversity conservation, cultural identity and other, less tangible benefits. Economic arguments have usually won out, and some conservationists and local communities have therefore been reluctant to consider compromise solutions – because the burden of compromise has seldom been evenly spread. This problem is all the more acute when dealing with degraded landscapes, because remnant forest resources are often assigned little value and economic interests have even less reason to compromise.

The double filter of FLR is therefore worth communicating to stakeholders, as it introduces the concept of scale into the search for compromise. The double-filter principle reflects an acknowledgement of the inevitability of some site-level specialization and trade-offs between economic, social and conservation values of the land. However, the principle also reflects the notion that these individual site-level trade-offs must be balanced at the landscape level. This means, for example, that a private-sector company signing up to

FLR will be obliged not only to avoid filling the landscape with monoculture plantations but also to ensure that the plantations are configured in a way that has no detrimental impact on the supply of other forest goods and services at the landscape level. In return, the plantation company gains greater acceptance of their activities in the landscape by stakeholder groups that might otherwise have opposed their presence. The Thailand case study in Box 2.1 is a good example of how the double-filter principle can lead to equitable compromise solutions and resolve difficult conflicts.

Another benefit of the double filter is that it accommodates and encourages adaptive management. One problem in trying to achieve balanced outcomes through the restoration of degraded forest landscapes is that land-use policies, markets, stakeholder groups and even ecosystems change over time. As the ITTO restoration guidelines note:

> *The social and economic conditions that exist when a forest crop is harvested are seldom the same as those prevailing when a tree seedling first takes root, nor do the priorities of individuals remain the same. Strategies for the restoration, management and rehabilitation of degraded and secondary forests must adopt a long-term perspective, anticipating, as far as possible, future trends. But they must also be flexible and capable of adaptation to changing circumstances.*
> (ITTO, 2002)

The double filter helps the practitioner and other stakeholders meet this challenge by not obstructing changes to site-level management practices or land uses as long as human well-being and ecological integrity are not compromised at the landscape level. It can also help persuade some stakeholders to undertake site-level activities that contribute towards improved landscape-level functionality, as illustrated in Box 3.1.

In addition to using the double filter as the basis of arguments for the value and feasibility of FLR, practitioners will need to put in place robust and measurable indicators so they can demonstrate tangible progress over time.

The contribution of FLR to poverty reduction

In the past few years, many donors and governments have placed increasing emphasis on the development and implementation of poverty reduction strategies. Bilateral and multilateral donors are moving away from project-based disbursement to channel aid assistance directly to national treasuries. More importantly, ministries of finance now privilege those ministries and government agencies that they believe can most effectively contribute to rapid poverty reduction through improvements in health, education or economic opportunities. These public bodies receive a larger allocation of the national budget, while other bodies that are considered of marginal importance to national poverty reduction strategies are the first in line for budget cuts. Forest departments have tended to fall into the latter category.

Box 3.1 The double filter as the basis for adaptive management in dynamic landscapes

In the 1990s there was a great deal of interest in the observed regeneration of large tracts of tropical secondary forest on farmland in some Central American countries. There was much speculation as to how long it would take for this new forest to attain the attributes of relatively undisturbed forest cover and whether new laws were necessary to ensure this would happen. However, for at least some owners of these new forests the reality on the ground was a different matter. While the area of secondary forest showed an increase at a national level, this did not mean that farmers were permanently dedicating parts of their land to forests. Instead, farmers rested land for 10 or maybe even 15 years to maintain a land-bank that would supply fencing poles and other useful products, and they rotated the location of these areas over time. Efforts to regulate this activity by prohibiting secondary forest clearance could have been counterproductive and might have resulted in less secondary forest being maintained on farms, with negative consequences for the overall integrity of the landscape.

How then to persuade senior government officials that the restoration of forest landscapes is worthy of support under such difficult funding circumstances? Somewhat paradoxically, poor people rely more on natural resources, particularly degraded natural resources, than other sectors of the population, even though they are often denied formal permission to utilize such resources. Experience has shown that when poor people are given long-term secure rights over degraded forest resources and supported with good technical advice they can turn such resources into healthy, productive and biologically rich assets within a few years. An economist might question whether this is enough by itself to lift poor people out of poverty. It seldom is, but it does constitute an effective and efficient first step, particularly in rural areas, where up to 75 per cent of very poor people live. Degraded and secondary forest resources are assets that can be deployed today – they already exist *in situ* – and can therefore help people start their transition out of poverty until other assets such as clinics, schools and new enterprises are commissioned and functioning. And healthier landscapes can contribute more generally to a country's poverty reduction strategy by enhancing the environmental security of poor people living or working in vulnerable locations.

Any plans to empower poor people to take decisions on and benefit from natural resource management need to be followed up with immediate action. Two key lessons that have been learned on devolving management and decision-making to local people are:

1   Do not promise what cannot be delivered. Ensure that you have sufficient authority to follow through with agreements and that communities do not find themselves in a situation where they lose control over the resource just as it becomes productive.

2   Focus on lifting constraints to local management rather than putting in place additional ones. Management rules should be limited and clear about what they prohibit. Detailed rules about how people should carry out silvicultural operations are rarely appropriate.

Contribution of FLR to local economic growth

Strong economic growth is a high priority for political decision-makers and is regarded as one of the principal tools for lifting large numbers of poor people out of poverty in a relatively short period of time. Economic planners and treasury officials spend a good deal of time considering how to make macroeconomic conditions more conducive to stimulating economic growth. At first glance such concerns may seem completely unrelated to forest conservation and FLR, and it is true that the forest sector (especially where forests are degraded) can never be expected to make the same contribution to national economies as many other sectors. However, the forest sector still has a role to play, particularly in stimulating local economic growth in places that have not, or will not, benefit from the trickle-down effects of globalization and national-level growth.

The benefits of national economic growth are seldom distributed evenly across all sections of society. In general, countries experiencing high economic growth rates (such as China) are also seeing a widening in the gap between rich and poor. What can be done to stimulate economic growth in poor rural areas? Part of the answer has been provided in the previous section – permit people to invest, use and enhance the productivity of degraded and secondary forest resources. Box 3.2 contains two case studies of how FLR has helped create new assets that contribute directly to local economic growth.

Two other points from the case studies in Box 3.2 are worth highlighting. The first is that the double filter of enhanced human well-being and restored ecological integrity means that the generation of landscape-level benefits (including tradable commodities, roundwood, non-timber forest products and environmental services such as carbon sequestration and storage, and improved water quality) does not prevent site-level specialization (such as plantations of small woodlots). The second point is that local communities and smallholders require long-term security if they are to benefit from degraded and secondary forest resources. The same argument of course also applies to large-scale owners and concessionaires, but their rights tend to be much more secure.

Box 3.2 Contributions of FLR to local economic growth: Two case studies

In the Sukhomajri watershed, India, joint forest management arrangements with local communities resulted in a 100-fold increase in the tree density of native species on denuded slopes over a period of 16 years. Subsequent increases in the production of forest grasses led to a six-fold increase in milk production, while better-regulated water flow permitted local investment in more diverse and higher-yielding cropping systems. As a direct result of this increased economic activity, household incomes across all social classes increased by 50 per cent. Further downstream, the siltation rate of an important lake near the major city of Chandigarh was reduced by 95 per cent, saving the city US$200,000 annually in dredging costs (Kerr, 2002).

In Chiapas, southern Mexico, poor farmers have established a global enterprise, selling high-quality carbon offsets to polluting businesses in the developed world. Since 1996, over 700 farmers have joined the Scolel Té initiative, planting native pines, cedars and fruit trees on their own farmland in configurations of their own choosing. Two-thirds of the income generated goes straight to the farmers, providing them with investment capital worth US$800 per hectare to help restore productive forest and agroforestry systems on currently degraded sites. In addition to supplying an emerging international market for carbon offsets, these homestead and farm plantations have also increased the supply of locally tradable commodities such as timber, fruit, medical plants and fuelwood.

Contribution of FLR to environmental security

Adequate environmental security for people living in vulnerable locations such as upland valleys requires not only that they are relatively well-protected from the impacts of catastrophic events such as periodic flooding but also that they have access to contingency livelihood alternatives when disasters do strike. Environmental security is an issue of increasing importance as more and more people are living in high-risk areas and/or relying on degraded resources to sustain their livelihoods. This means that more people are likely to be affected when a flood, drought or landslide occurs. Further, while environmental security is often talked about in terms of catastrophic natural events, its principles are equally relevant in times of major economic shock or violent conflict. Decision-makers also need to be reminded that environmental security can sometimes be an issue of national security, as the hardship brought about by natural disasters can lead to increased civil unrest.

The links between FLR and environmental security are relatively straightforward. Loss of forest functionality in degraded landscapes has both *in situ* and downstream impacts. For example, as forest land is degraded and fragmented, the velocity and rate of site-level run-off increases, soil erosion accelerates, slope stability is reduced, siltation loads increase and water quality declines. The disasters that grab headlines are therefore not just a consequence of, for example, one particularly heavy rainfall but are symptomatic of a long-term erosion of ecological integrity. FLR can help reverse this trend not only by increasing landscape-level resilience to shocks but also by enhancing landscape-level adaptability so that both government and local communities are better placed to respond to such shocks.

Contribution of FLR to biodiversity conservation

Over 12 per cent of the earth's forest cover is now under some form of protected area status consistent with the IUCN categories. Although this is an admirable achievement, much remains to be done to improve the management of these areas – and even then it is becoming clear that these areas may be insufficient to safeguard forest biodiversity. One option is to continue to put more land under protection; this is certainly needed for forest types that are particularly threatened and poorly represented within protected area systems and for areas that still have large tracts of intact biologically rich forest. Yet the reality in many tropical countries is that these conditions no longer prevail, and local communities are becoming increasingly vocal about losing their land rights, whether for protection or production purposes. There is now growing awareness that protected areas, no matter how well-managed, will continue to face major challenges if they are simply 'islands of conservation' in an otherwise hostile sea of unsympathetic land uses. This leaves governments with a dilemma, given that many of them are committed to the Convention on Biological Diversity's target of halting and reversing biodiversity loss by 2010.

While FLR can never substitute for a representative network of protected areas, it can certainly contribute to the ability of these areas to conserve biodiversity. The case study in Box 2.2 showed the critical role that a technical package of ecological restoration and planted production played in enhancing the viability of protected areas in one particularly 'hostile' landscape. And Figures 2.2a and 2.2b demonstrated the remarkable ability of threatened species to recolonize even badly degraded sites, provided the right conditions prevail. FLR offers one of the most promising options to ensure that environmental values are represented outside protected areas in a way that accommodates other land-use requirements and enhances the conservation value and ecological integrity of the protected areas themselves.

It could be argued that the rationale given here for FLR is simply a reworking of the justification used to promote integrated conservation and development projects (ICDPs). Given that these projects have produced very mixed results, one could question whether FLR will fare any better. However, while FLR has certainly been influenced by ICDPs, there are also a number of critical differences. Many forest-based ICDPs were established in relatively intact forest areas, a proportion of which had been, or would soon be, gazetted as protected areas (often entailing a loss of forest-use rights for the local communities). Part of the rationale of ICDPs was that they would compensate local communities for lost livelihood opportunities through the provision of alternative development projects. The context for FLR is different in two ways. First, the condition of the forest resource outside the protected area will, by definition, already be heavily modified or degraded. Second, the process of restoring the productivity and ecological integrity of degraded resources is likely to entail a transfer of additional rights (and responsibilities) to communities or individuals.

Conclusions

Forests now contribute less to many countries' GDP than they used to and are often no longer seen as a resource of national strategic importance. Financial support for forests has been reduced, both from national budgets and as a percentage of donor aid flows. Put simply, decision-makers in many countries do not consider forests to be as important as they once did. While many of us may see this as short-sighted, given the wide range of goods and services that flow from forests and the direct and indirect contribution that forests can make to national development, we, as foresters, must also accept part of the blame for this situation. For too long, one single forest good – industrial roundwood – was promoted to the near exclusion of all other goods and services.

If we are to turn this situation around then we must begin to make a more convincing case. FLR provides one possible vehicle for doing so, because it promotes all the values that forests can provide to society: hydrological regulation and watershed protection, carbon sinks, an on-farm asset, soil stabilization, an insurance policy in times of natural disasters, and a repository of biological diversity of both local and global value. However, such arguments will not stand by themselves – they must be presented in terms that decision-makers clearly understand. If finance ministries are concerned with poverty reduction strategies, then forest managers must learn how to make a case in these terms – not simply with respect to maintaining subsistence livelihoods but also to contributing to local economic growth.

References and further reading

ITTO (2002) *ITTO Guidelines for the Restoration, Management and Rehabilitation of Degraded and Secondary Tropical Forests*, ITTO Policy Development Series No 13, ITTO, Yokohama, Japan

Kerr, J. (2002) 'Sharing the benefits of watershed management in Sukhomajri, India', in S. Pagiola, J. Bishop and M. Landell-Mills (eds) *Selling Forest Environmental Services – Market-Based Mechanisms for Conservation and Development*, Earthscan, London, UK

4

Applying an Adaptive Management Approach in FLR

Don Gilmour

This chapter sets out what is involved in managing an FLR initiative and proposes the adoption of an adaptive management approach to enable FLR practitioners to respond to the dynamics found in natural and socio-economic systems.

Management characteristics of FLR initiatives

FLR initiatives typically have several characteristics that make them quite different from conventional forestry operations. These include:

- multiple stakeholders with multiple interests (local, regional and national);
- complex ecological systems across a large landscape, with a variety of land uses;
- the interface between large-scale natural systems and social systems; and
- a high level of uncertainty and many unknown factors.

A consequence of this combination of factors is that FLR managers need to adopt a social learning approach based on a process of experiential decision-making and monitoring. Adaptive management has been developed to accommodate these characteristics. It is an approach to the management of complex systems based on incremental, experiential learning and decision-making, supported by active ongoing monitoring of and feedback from the effects of outcomes of decisions (Box 4.1). The approach has elements of trial and error, but it is much more than this, incorporating explicit learning as part of a process of building social capital among multiple stakeholders. It therefore involves:

- collaboration and learning;
- combining the learning and action that take place within a group of people (capturing both knowledge generation and the application of this knowledge in action); and
- knowledge-sharing among group members.

Box 4.1 Origins of adaptive management

The term 'adaptive management' was coined in 1978 by an interdisciplinary team of biologists and systems analysts to describe a guiding principle for managing the interface between society and the biosphere. It was based on detailed studies of complex ecosystems, such as the Florida Everglades, the Columbia river and the Baltic Sea, in which humans play a dominant role. Adaptive management has since become a major approach, informing real attempts to manage large ecosystems in a sustainable manner.

Adaptive management offers three important benefits:

1 it can avert crises in conditions of uncertainty and surprise by increasing the societal capacity to 'roll with the punches';
2 it offers a social steering instrument that can complement market, fiscal, regulatory and other measures to strengthen broad-based, multi-stakeholder engagement in the evolution of more sustainable relations between people and their environment; and
3 it offers a way in which scientific-based technologies, alongside an understanding of people's perspectives, values and meanings, can contribute to collective learning and motivation for action.

Key components of adaptive management

It is convenient to think of adaptive management as a series of interrelated processes:

- ***understanding the social and biophysical context*** at multiple levels; this involves identifying stakeholders and dealing with multiple (and sometimes conflicting) interests;
- ***negotiating objectives and outcomes*** for different levels;
- ***applying action learning*** (plan, act, observe and reflect) to facilitate the implementation process; and
- ***monitoring and impact assessment***.

These processes should not be thought of as a series of sequential steps in which you complete one management task before moving on to the next. Rather, the processes should be thought of as interrelated and overlapping. For example, collecting and updating information to understand the context will be a process that continues throughout the life of the initiative. Likewise, monitoring and impact assessment is not just a one-off activity at the end of the initiative but an ongoing practice that feeds constantly into the action learning cycle from the very beginning of the intervention.

Each of the four key components of adaptive management is now considered in turn.

Understanding the context

The context of an FLR initiative comprises the social and biophysical conditions in which it takes place and which could have an impact on it (Table 4.1). While it is never possible to understand everything about the context (particularly as the context will change over time), it is important to know enough about it to make a start. An improved understanding of the context can be gained while the initiative continues; an action learning approach recognizes that updating knowledge of the context is an important part of the management process.

Negotiating objectives and outcomes

The objective of an FLR initiative will vary depending on the agenda of the group promoting the landscape restoration. A forest department, for example, might want to restore an area of degraded forest land primarily to improve timber production, while a conservation agency or NGO might want to improve

Table 4.1 *Examples of the context of an FLR initiative*

Biophysical	• Type, condition and location of forest patches
	• Type and location of non-forest land
	• Presence or absence of degrading influences
	• Trends in forest condition – for example increase or decrease in forest areas
	• Drainage pattern and slope characteristics
	• Land-tenure patterns (legal and *de facto*)
	• Geological and soil patterns
Social	• Location of settlements
	• Dependence of local people on forest resources for livelihood support
	• Existence of local social institutions (including NGOs)
	• Conflicts over land or resource use
	• Stakeholder groups (inside and outside the landscape) that have an interest in the FLR initiative

Box 4.2 The action learning cycle

Step 1: Plan

The action learning cycle starts with planning to take action on some pre-defined issue or problem situation. The planning is built on the experience and ideas of all partners involved, because learning is enhanced when it is derived from day-to-day work and experience.

Step 2: Act

The results of the planning are put into practice, using time-frames agreed in the planning sessions.

Step 3: Observe and reflect

Those involved observe the results of the action and reflect on the impact. Reflections need to be carried out explicitly and are best done as a group, ideally facilitated by an outsider in the early stages. A series of questions can help to focus the reflection. Suitable questions could include:

- What changes have taken place?
- What were the strengths and weaknesses of what was done?
- What could have been done better?
- What problems have resulted from the changes?
- Were there unintended consequences of the actions?

This reflection is very important as it enables the next steps in the cycle to benefit from the explicit learning that has resulted from the previous action.

Step 4: Draw lessons

Lessons are drawn from the previous steps of action and reflection. The experiences to date are linked back to the concepts and ideas that were used in the initial planning. This leads to replanning for the next cycle, building on the learning of the various steps of action and reflection on and drawing lessons from previous cycles. In this way, planning and action can proceed incrementally, with everyone participating in and contributing to all facets of the process. Thus there will be a strong sense of ownership over the outcomes (both successes and failures).

habitat for wildlife or restore an endangered biotype. Hence the primary objective of the group initiating the rehabilitation or restoration activity may create different responses from different stakeholders. It is only by identifying the interests of the various stakeholder groups that negotiations can occur, and the initial objectives may need to be modified to take account of the interests

---

Box 4.3 What is meant by uncertainty and risk?

It is never possible to control all the variables when managing initiatives, such as FLR, that involve multiple stakeholders, multiple interests and complex landscape-level issues. There is always incomplete or imperfect knowledge and therefore much uncertainty. Factors may arise that were not known or not considered important at the outset and these may influence the outcome in ways that were not planned or expected (see example in Box 4.4).

Forest activities contain much greater inherent levels of risk than most agricultural activities, largely because of the long-term nature of forest crops. The most common risks include market shocks, disease and fire.

---

of other stakeholders. This process inevitably involves trade-offs and requires compromises in order to achieve outcomes that will be socially acceptable and sustainable over the long term.

Applying action learning

The key idea behind action learning is that a group of people with a shared issue or concern collaboratively, systematically and deliberately plan, implement and evaluate actions (Box 4.2). It is a process of learning through experience in order to act more effectively in a particular situation and is well-suited to situations with a great deal of uncertainty and risk (Box 4.3).

The process should be thought of as ongoing rather than as a one-off event (as illustrated in Figure 4.1). The participants continually go through the

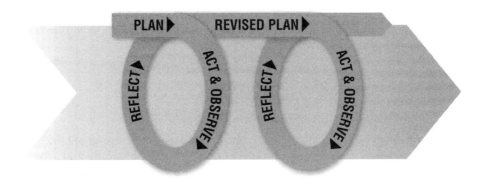

Figure 4.1 *The action learning spiral*

Source: Kemmis and McTaggart (1988)

Box 4.4 Monitoring for action learning: Case study from Nepal

An attempt to rehabilitate the degraded hillsides of common land in a region of eastern Nepal was eagerly accepted by local people, as evidenced by discussions at village meetings. However, after the first year of planting it was noted that most of the planted trees had not survived. Discussions with a wide range of local people, outside a formal meeting setting, revealed that a group of poorer people (who were not sufficiently empowered to speak at village meetings) disagreed with the rehabilitation proposal. Their livelihoods were largely dependent on managing herds of grazing animals and they did not wish to lose their grazing land. The low survival rate of the planted trees was due to the graziers having allowed their animals to graze the recently planted hillsides. Their more wealthy and powerful neighbours were primarily sedentary agriculturalists and did not need much open grazing land. This finding enabled the original approach to be modified so that the economic needs of the graziers were taken into account, resulting in greater success in the rehabilitation initiative.

The lessons from this example are that:

- ongoing monitoring enabled problems to be identified before they became too serious, so that the next action learning cycle could be adjusted based on the learning obtained in the previous cycle;
- even with what seems like thorough planning, there are almost always unexpected outcomes and unintended consequences that need to be explicitly looked for and learnt from before continuing with the next action learning cycle;
- great care needs to be taken to identify all the stakeholder groups that will have an interest in the outcomes of the rehabilitation or restoration activities; and
- consensus at village meetings does not necessarily mean agreement by all interest groups, particularly where there are large differences in power relations between different groups.

cycle, with each iteration improved by the knowledge and learning obtained in previous cycles. Within the broad framework of reflective planning and action, various methods and tools can be used to collect information. This process is sometimes called action research to emphasize the importance of researching or exploring new or innovative approaches to a problem. Thus it is not *just* about learning, although learning is one of the important outcomes of the process.

Monitoring and impact assessment

An ongoing approach to monitoring and impact assessment is an essential aspect of adaptive management. This enables the stakeholders to build their social capital by sharing the learning that comes from such assessments. The next action learning cycle of planning/acting/observing/reflecting is updated by realistic information, thus helping to maintain maximum adaptability and flexibility (see the example in Box 4.4).

In addition to ongoing monitoring and assessment throughout the implementation process, there needs to be monitoring of the 'big picture' aspects of the programme, based on the overall objectives. This is best done by establishing indicators by which results can be judged. An example from a WWF FLR programme in New Caledonia is given in Table 4.2. It should be noted that this example focuses on the biophysical aspects of FLR and does not include any socio-economic aspects.

Table 4.2 *Monitoring indicators developed for
an FLR programme in New Caledonia*

Operation phase	Indicators
Acquiring better knowledge	• Animal and plant species studied (seed sources, nursery techniques, etc.) • Scientific reports published • Surface area covered by census
Ensuring protection	• Sites and areas protected • Lengths of fencing and fire barriers installed • New texts, regulations and procedures adopted • Surface area of burnt and damaged forest reduced • Boundaries listed
Rehabilitation	• Forestry and hunting management plans • Surface area and species planted and maintained • Species planted and cultivated in nurseries • Rare species saved from extinction • Level of invasion by unwanted animals and plants reduced
Valorization	• Substances and plants newly marketed • Brochures, flyers, posters and signs created and distributed • Events (exhibitions, seminars, etc.) carried out • People contacted (schools, local residents) • Press and magazine articles published • Sites and discovery trails laid out
Sustainable management	• General and thematic maps drawn up • Percentage of dry forests managed, planned and protected • Cooperation agreements signed with landowners

Source: Adapted from WWF (no date)

Table 4.3 *Guide to using an adaptive management approach for an FLR initiative*

Component	Description
Understanding the context (see Box 4.1)	• Identify the key stakeholders • Understand the policy context • Understand the socio-economic conditions of people living in the landscape • Understand the biophysical and forest management context • Understand institutional settings (formal and informal, government and non-government)
Negotiating objectives and outcomes	• Develop a shared vision for the future (including objectives and outcomes – both biophysical and socio-economic) • Develop indicators of successful FLR (biophysical and socio-economic) • Assess current conditions against the ideal conditions agreed in the vision • Identify and prioritize critical areas and actions needed to achieve objectives
Applying action learning to facilitate the process (see also Box 4.2)	• Establish action learning groups and introduce the action learning processes • Plan for action • Action • Observe, monitor and reflect on the results of the action • Draw lessons from the outcomes to improve further actions • Continuously improve the management strategies through application of the action learning cycle
Assessing the impact	• Assess the impacts of FLR activities on the biophysical conditions across the landscape and the socio-economic conditions of the key stakeholder groups, paying particular attention to those parts of the process around which there is greatest uncertainty

Implementing adaptive management of an FLR initiative

Table 4.3 outlines the key components of the adaptive management approach and sets out some examples of what is involved in each component when applying the approach in an FLR initiative. It must be stressed again that breaking down the implementation process of adaptive management into a series of discrete steps is somewhat problematic because it might give the impression that adaptive management is a simple linear process, akin to applying a prescription or blueprint. Here it is worth recalling the cautionary comments in Box 4.3 on the social and biophysical uncertainties of forest management. It is better to think of the adaptive management process as a series of action learning loops rather than a straight line from planning to the achievement of planned outcomes. Managers should feel free to adapt and modify the approach based on the learning that comes from the application of action learning throughout the process.

References and further reading

Adnan, H. (2005) *Learning to Adapt: Managing Forests Together in Indonesia*, CIFOR, Bogor, Indonesia

CIFOR Adaptive Collaborative Management website: www.cifor.cgiar.org/acm

Hartanto, H., Lorenzo, M. C. B., Valmores, C., Arda-Minas, L., Burton, L. and Prabhu, R. (2003) *Learning Together: Responding to Change and Complexity to Improve Community Forests in the Philippines*, CIFOR, Bogor, Indonesia. An Indonesia-focused version of this publication is also available: Kusumanto, T., Yuliani, L., Macoun, P., Indriatmoko, Y., Fisher, R. and Jackson, W. (1998) 'Action research for collaborative management of protected areas', Workshop on Collaborative Management of Protected Areas in the Asian Region, Sauraha, Nepal

Kemmis, S. and McTaggart, R. (eds) (1988) *The Action Research Planner*, 3rd edition, Deakin University Press, Geelong, Australia

Ogilthorpe, J. (ed) (no date) *Adaptive Management: From Theory to Practice*, SUI Technical Series, vol 3, IUCN, US Office, Washington, DC, US

Weinstein, K. (1999) *Action Learning: A Practical Guide*, Gower, Hampshire, UK

WWF (no date) *The Dry Forests of New Caledonia*, WWF, Noumea, New Caledonia

5

Understanding the Landscape Mosaic

Don Gilmour

Introduction

This chapter shows how a landscape can be thought of as a mosaic (or patchwork) made up of different components (for example, various land uses, land tenure, drainage patterns and human settlements) and suggests some tools that can be used to understand and represent this. The chapter also highlights the importance of considering the contributions of different parts of the landscape (including both forest and non-forest areas) to the overall objectives of an FLR initiative.

What is a landscape mosaic?

A landscape mosaic is made up of different components, pieced together to form an overall landscape-level 'patchwork'. The actual composition of the mosaic and the pattern in which the components are distributed will be unique to each landscape.

A landscape mosaic can be represented in a variety of ways using, for example, maps, tables of different attributes and written descriptions. Maps are particularly useful as they can present complex situations in a visual manner and can help achieve a common understanding among different stakeholder groups; Figure 5.1 shows a hypothetical landscape mosaic map. It is much easier to visualize what is being discussed when using maps than when the information is presented as lengthy written descriptions or tables. Maps can also be used to cross-check information from different sources. For example, official records of land tenure can be mapped and shown to local stakeholders to discuss any differences with their own understanding of land tenure, including access and use rights. Maps are also very useful when discussing issues such as landscape-level objectives, restoration and rehabilitation options, and trade-offs between

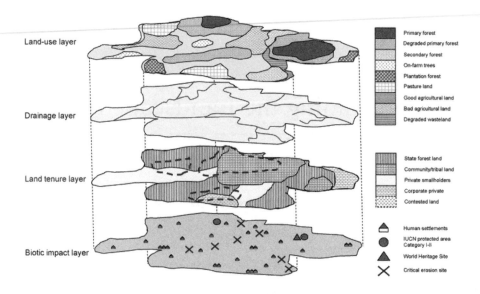

Figure 5.1 *Layers of a forest landscape mosaic map (hypothetical example)*

different objectives and stakeholder groups. Because of the complex nature of most forest landscapes, it is usually necessary to use several maps, preferably of the same scale, that can be overlaid on each other to build up a composite picture (as in Figure 5.1).

Key components of a landscape mosaic

While a certain minimum amount of information is required before effective planning and implementation can commence, there is often a tendency to spend an inordinate amount of time and effort collecting information simply because it is available. It is generally better to commence with a minimum information base and build on it as the need arises – using the 'action learning cycle' to reflect on the situation during the implementation process (see Chapter 4).

It is worth posing the question: What is the minimum information needed before an FLR initiative can commence? The answer will of course depend on the situation, but the most common types of information required are outlined in Table 5.1, along with suggested sources of such information.

Tools and techniques to map and describe the landscape mosaic

As indicated above, maps are one of the most useful tools for presenting information on the landscape mosaic in a format that is meaningful to the

Table 5.1 *Information on key components of the landscape mosaic needed for planning FLR strategies and activities*

Key components of the landscape mosaic	Uses of information	Sources of information
Land use		
Land-use patterns (different categories of forest, agricultural and pastoral land). See land-use overlay in Figure 5.1.	Strategic planning purposes.	Maps, aerial photographs.
Trends in land use (e.g. forest areas increasing or decreasing; forests becoming more or less degraded; agricultural areas increasing or decreasing). Different stakeholders may have different views on these trends.	Determining overall restoration and rehabilitation strategies.	Discussions with key informants, government officials, local farmers, scientists, etc. Remember that local views can differ from official views, and cross-checking may be necessary.
Population patterns and labour availability.	For example, identifying spare time in the agricultural calendar that could support restoration and rehabilitation activities.	Official records; discussions with key informants, particularly local people.
Local (indigenous) knowledge of history, harvesting practices, ecological aspects, ethnobotany.	Cross-checking information derived from official sources and informing restoration and rehabilitation strategies.	Discussions with local communities and researchers who have worked in the area.
Drainage		
Physical landscape features (e.g. contours, streams, drainage lines). See drainage overlay in Figure 5.1.	Planning restoration and rehabilitation strategies.	Maps, aerial photographs.

Table 5.1 *continued*

Key components of the landscape mosaic	Uses of information	Sources of information
Land tenure		
Land ownership. See land-tenure overlay in Figure 5.1.	For example, identifying key stakeholders.	Cadastral boundaries will give official legal situation. Discussions with land occupiers or managers will give local views of use rights, which could differ from the official view.
Historical legacy of different or contested tenure (or access and use rights).	Determining restoration and rehabilitation strategies that will be sustainable.	Official records; discussions with government officials, NGOs and local people (again, remember that official perceptions may differ from local ones).
Biotic impact		
Where are the problems: threatened species, biodiversity hotspots, eroding areas, fragmented habitats, weeds or pests? See biotic impact overlay in Figure 5.1.	Determining restoration and rehabilitation strategies.	Maps, aerial photographs, publications, local knowledge, specialist knowledge (government and NGO scientists, etc.).
Others		
Infrastructure (including roads, railways, towns and villages).	General planning purposes.	Maps, aerial photographs.
Geology and soil types.	Deciding, for example, on appropriate species for planting in different sites.	Maps and local knowledge.

majority of stakeholders. However, maps will usually need to be supported by quantified information to assist with planning field activities and monitoring. Government sources of information are normally readily available, but information from these sources often needs to be cross-checked with locally available data and perceptions. Maps can be out-of-date or simply wrong. And just because quantified data is collated in official documents does not necessarily mean that it is correct – there is an old adage that cautions against confusing numbers with facts!

Most countries have geographic information systems (GIS) available in central land management agencies, and these can be a rich source of information for building up the components of a landscape mosaic. It is generally possible to access these databases to prepare base maps for landscape-level initiatives such as FLR. Tools such as the computer Mapmaker program are readily available and are well-suited for use at the local operational level.

Contribution of key areas of the landscape to FLR initiatives

Different parts of the landscape need to be assessed to determine how they can contribute to the overall FLR objectives, as outlined in Table 5.2.[1]

Caution is required in determining the real availability of forest and non-forest areas for restoration or rehabilitation. In some cases, availability is perceived differently by different stakeholder groups. In many situations, for example, much non-agricultural land may be under some form of common property management regime that is well understood by local communities but contested by government officials. Much of the 'degraded' forest land under fallow as part of shifting cultivation cycles would fall into this category. It is important to identify and resolve these different perceptions of availability before starting the restoration or rehabilitation activities.[2]

Contribution of restored landscapes to conservation and development objectives

One of the factors that works against FLR is the belief of some land-use planners and managers that commercial benefits are maximized by replacing natural forests with high-value cash crops or fast-growing tree plantations. Such approaches lead to a simplification of the landscape, a reduction in its ecological functions and, possibly, a decline in agricultural or forest productivity. Ensuring conservation and maintaining productivity requires more than just the setting aside of small protected areas on land unwanted for production. Instead, conservation and production objectives require that sufficient biodiversity is retained across the landscape to maintain key ecological processes, such as nutrient and hydrological cycles. Intensive land-use activities

Table 5.2 *Contribution of key areas of the landscape to an FLR initiative*

Key areas of the landscape	Contribution to an FLR initiative
Forest areas	
Intact natural forest (large areas)	These contain much of the conservation and development values of the initial forest landscape and are often the key building blocks for FLR initiatives. They generally need to be connected with restored and rehabilitated areas of the landscape to strengthen their contribution to FLR objectives.
Intact natural forest (small areas)	These provide important conservation and development values on site that can be enhanced by expansion and connection to other key forest patches and areas to be restored and rehabilitated.
Plantations	These contain some conservation and development attributes that can be enhanced by management. They can also serve as useful buffers around degraded forests and protected areas.
Degraded forest or shrublands (large areas)	These can be key targets for restoration and rehabilitation and for connecting to other parts of the forest landscape.
Degraded forest or shrublands (small areas)	These can provide some conservation and development values that can be enhanced by restoration and rehabilitation and by connecting these areas to other key parts of the forest landscape.
Non-forest areas	
Farmland	Management of this land can be modified to contribute to FLR objectives (see example in Box 5.2).
Trees on farms	These can contribute to conservation and development outcomes, particularly if connected with intact forest patches.
Riverine (riparian) strips	These are important habitat types and building blocks for connectivity in the landscape. They may require restoration or rehabilitation to protect both on-site and downstream soil and water values.
Degraded areas	These provide an opportunity for rehabilitation for on-site conservation and development benefits and for improved connectivity between natural forest patches.
Eroded areas, landslips	These require special treatment to protect both on-site and downstream values.

are often appropriate but need to be undertaken within a landscape that retains its ecological functionality – in other words one that is biologically diverse and spatially complex. Any simplification of the landscape mosaic, by, for example, replacing natural forest patches with industrial timber plantations, could reduce the capacity of the remaining landscape components to maintain biophysical and socio-economic outcomes. FLR is a means of halting the degradation and building on what remains in order to restore landscape functions (such as hydrological processes).

One of the underlying arguments behind the promotion of FLR is that the effectiveness of efforts to conserve biodiversity and restore key functions at particular sites depends on these restored areas complementing the existing mosaic such that the whole is more than the sum of the parts. For example, the conservation value of a small patch of isolated remnant forest will be largely limited to the on-site value. Similarly, the accumulated value of many such small fragmented patches will also be limited. However, once a patch of forest is connected to other patches by creating a corridor or restoring adjacent degraded land, its value, both on and off site, will be greatly increased in terms of biodiversity conservation and the maintenance of ecological functionality. This in turn can lead to overall improvements in agricultural productivity.

Contribution of landscape components to FLR objectives

The way in which various components of the landscape will be treated depends on the objectives of the FLR activity that are agreed on by the key stakeholders.[3] As discussed in Chapter 4, a key determinant of the FLR objectives is likely to be the agenda of the group that is initiating the landscape restoration programme. For example, a forest department might want to rehabilitate an area of degraded forest land primarily to improve timber production, whereas a conservation agency or NGO might want to rehabilitate land to improve habitat for wildlife or to restore an endangered biotype. In most cases the trade-offs between conservation and development outcomes will require considerable negotiation between stakeholders.

The following discussion gives several examples of the way in which different components of the landscape can be managed to contribute to FLR objectives.

Plantations, which are normally established primarily for timber production, can also be managed to provide significant conservation and development values, over and above any conservation and development objectives originally considered. In many parts of the world indigenous species develop as an understorey in monoculture plantations. Protection of the understorey can enhance plant and animal biodiversity (a conservation benefit) and can often provide products of value to local communities (a development benefit), as illustrated by the case described in Box 5.1.

Box 5.1 Natural regeneration under monoculture plantations in Nepal

Degraded hillsides in the Middle Hills of Nepal are difficult to rehabilitate with the broadleaf forest species which originally grew on the sites and which are preferred by local communities. The best results have come from using chir pine (*Pinus roxburghii*) as the major plantation species, because of its pioneer attributes. There was serious debate about the selection of species for rehabilitation and substantial criticism of the use of chir pine. However, the choice had to be made between using locally desirable broadleaf species (which were ecologically unsuitable under the prevailing degraded conditions) and the less-preferred chir pine, which could survive on the degraded sites.[4] It was subsequently found that if the plantations are protected from grazing, an understorey of broadleaf species often appears five to ten years after initial establishment, as environmental conditions change. The understorey species (which can be at densities of more than 2000 stems per hectare) include many that are valued as fodder and firewood and for medicinal purposes. Future management options can involve a mix of strategies depending on the objectives of the key stakeholders.

It is worth noting that the development of high-value understorey species in the case described in Box 5.1 was an unexpected result and came to light through the application of action learning approaches and adaptive management.[5] Reflection on these results led to significant changes in approaches to plantation management in parts of Nepal, and consequently a wider range of benefits (in terms of both conservation and development) were made available to both the local and wider communities.

The case study in Box 5.1 also illustrates how rehabilitation can start with a relatively low-cost and low-technology approach (a single-species mono-culture) but lead to a multi-species forest with multiple conservation and development benefits.

Plantations can also serve as buffers around restored and protected areas. In some settings they also effectively mark the boundaries of 'managed land', so that local people know to keep grazing animals out and to refrain from burning.

Conservation and development objectives can also be enhanced by linking restored and rehabilitated parts of the landscape with areas that are already in good condition and under a high level of protection. Areas can be selected for rehabilitation and restoration so that they act as corridors to connect protected areas to other parts of the landscape. Similarly, areas can be selected for rehabilitation and restoration so that they act as buffers for protected areas to cushion them from degrading influences such as grazing and fire. Strategic actions such as these will greatly increase the overall impact of the rehabilitation and restoration activities, because the on-site impacts are multiplied greatly by the added benefits that are provided to adjacent areas.

Trees on farms can also contribute to both development and conservation objectives – to the former, for example, by increasing the availability of timber and fruit for local use and sale and to the latter by increasing plant biodiversity and improving wildlife habitat. In addition, the conservation value of trees on farms can be greatly increased if they are connected to other patches of forest in the landscape. Slight modifications to existing farm practices can often yield both conservation and development benefits at minimal cost (see the example in Box 5.2).

Box 5.2 Restoration of agricultural land in Australia

Grass verges around sugarcane fields in tropical northern Australia provided habitat for rats, causing major losses of cane. Tree-planting along these verges has greatly reduced the tall grasses and consequently eliminated the rat problem, thus providing the dual benefits of improved habitat for natural species and improved agricultural productivity. Ensuring that the trees planted are local indigenous species will maximize the conservation benefits.

Even if they are degraded, remnant patches of forest can contribute towards FLR objectives, particularly if they are connected by corridors. It might be more cost-effective to spend a relatively small amount of money carrying out a few strategically important activities than to spend large sums of money on large-scale rehabilitation or restoration activities. Strategic activities could include, for example, the creation of a rehabilitated corridor to connect degraded forest patches and then the protection of these patches from degrading influences such as grazing and fire. Such low-cost and low-technology options could yield major benefits in the long term.

Notes

1   See Chapter 8 for more discussion on how different areas of the landscape can contribute to FLR initiatives.
2   See Box 4.4 for an example of different perspectives of availability.
3   See also the discussion on site-level options in Chapter 8.
4   See Chapter 13 for more on trade-offs.
5   See Chapter 4 for more details on both action learning and adaptive management.

References and further reading

Bennett, A. (1999) *Linkages in the Landscape: The Role of Corridors and Connectivity in Wildlife Conservation*, IUCN, Gland, Switzerland and Cambridge, UK

Lamb, D. and Gilmour, D. (2003) *Rehabilitation and Restoration of Degraded Forests*, IUCN, Gland, Switzerland and Cambridge, UK, and WWF, Gland, Switzerland

Mapmaker program: www.mapmaker.com

6
Understanding Forest Landscape Dynamics

Wil de Jong

This chapter is based on the premise that the forest landscape we want to manage and restore is the product of dynamic forces operating as direct or indirect causes of change. The chapter highlights the importance of understanding and addressing the forces responsible for landscape change to ensure that restoration efforts are successful; it is closely linked to Chapter 5 on understanding the forest landscape mosaic.

Essentially, landscape dynamics include changes in the composition of the landscape (that is, the make-up of the various landscape components such as forest land, agricultural land or housing) and changes in the condition of individual components (such as conversion of agricultural land from grazing to crop production).

Direct causes of landscape dynamics might include:

- logging operations;
- conversion of forests for estate crops;
- forest encroachment by poor people searching for agricultural land;
- road construction;
- development and expansion of housing settlements; and
- industrial pollution of rivers.

Behind each of these direct causes is a series of underlying causes or drivers. For logging operations, for example, these drivers might include:

- changing international demand for timber;
- the exhaustion of timber reserves in other locations;
- changing prices or international demand for estate crops;
- national tax incentives;

- national development plans driven by loans from international banks;
- drought; or
- changes in land ownership elsewhere that encourage in-migration to forested regions and forest clearance for agriculture.

How and why do forest landscapes change?

Landscapes can change dramatically over time. A simplified example of a forest landscape dynamic model starts with a region almost fully covered with forest. The first changes are caused by migrating farmers who come to reside in isolated settlements. They practise swidden agriculture and maintain a diverse landscape of agricultural fields, fields under fallow, tree crops, mixed orchards or forest gardens, and forest reserves. When other actors move into the landscape, the forest landscape begins to change drastically. Outsiders may be landless migrants in search of a better livelihood, timber companies looking for timber, or entrepreneurs or companies looking for agricultural land. The direction of the landscape change will vary depending on who is the main actor. Logging companies will modify extensive areas of forests and logged-over forest is prone to fires, which often become an important direct cause of change in such landscapes. Small farmers may follow in the wake of logging companies and become the major actors affecting the forest landscape. When large numbers of migrants move into a new forested region, they are more likely to settle at the forest frontier and progressively convert forest land into a mixed agroforestry landscape similar to that of the original settlers. When entrepreneurs or estate-crop companies are the main actors, the landscape dynamic will be dominated by a large-scale conversion of forest into grazing lands or estate-crop plantations.

Table 6.1 summarizes a conceptual distinction of three stages of change in tropical forest landscapes, reflecting a trajectory from a relatively highly forested, unpopulated landscape to one with a high population density and much-reduced forest cover. The table also illustrates how the early stages of landscape change often coincide with modifications of other features of the forest landscape, including settlements, roads, property rights, social relations, and the existence and enforcement of legislation. The actual dynamics seen in forest landscapes will show considerable diversity; Box 6.1 provides two examples of how a forested region can change over time and how the underlying drivers of these changes can differ in different contexts.

Guidelines for the analysis of forest landscape dynamics

For management purposes, we need to know the current status of the forest landscape (the mosaic) and the dominant forces that influence its dynamics. In general, a landscape is likely to evolve towards more intensive land use, and

Table 6.1 *A simplified model of forest landscape dynamics*

Changing characteristics	Extensive-use stage	Intensive-exploitation stage	Forest-depleted stage
Land use	Swidden agriculture by indigenous and other local groups, largely for subsistence	Logging, estate crops and high-pressure swidden agriculture by companies, entrepreneurs and migrants for markets and subsistence	Small-scale forestry, stable agriculture, agroforestry and plantations by communities
Population density and resource levels	Low population density, low capital, plenty of natural resources	Medium population density, high capital, decreasing natural resources	High population pressure, high capital, low natural resources
Infrastructure	Water and animal transportation; distant markets	Some road networks and motorized transportation; market access improved	Full accessibility by roads and transportation opportunities; markets available
Land tenure and other legislation	Customary property rights; customary laws	Conflicting property claims; little enforcement of state-sanctioned property rights or rule of law	Clear property rights, although much land still held by government; greater presence of government and enforcement of rule of law
Policies	Largely unaffected by government policies; often forgotten by governments	Forest viewed as national wealth and expected to contribute to national development; marginalization of local resource needs	Higher environmental and social awareness, partly because of local civil action; reforestation and conservation beginning to be emphasized

Box 6.1 Tropical forest landscape dynamics in Bolivia and Vietnam

Forest landscape dynamics in Santa Cruz, Bolivia

The department of Santa Cruz accounts for 51 per cent of Bolivia's forest cover and in 2001 experienced 60 per cent of the country's forest clearance. Until 1950, the department had only 60,000 hectares of land under cultivation. In the 1950s, small- and large-scale farmers began agricultural expansion to supply national markets with corn and rice. And from the 1970s import-substitution policies promoted the expansion of cotton and sugarcane production through road construction, subsidized agricultural credits and supported agricultural prices. As a result, by 1985 smallholders held 150,000 hectares of cultivated land while larger farms cultivated about 170,000 hectares. In the mid-1980s the government started to promote export-oriented agricultural production for regional markets, mainly soybeans, but it also slashed all subsidies and stopped other agricultural development support, except the granting of legal tenure of forest land to medium- and large-scale farmers. Large-scale producers became the main actors in the department's forest landscape dynamics; by 2000, small farmers cultivated 200,000 hectares and large farmers 700,000 hectares.

Within the last decade, however, indigenous groups in Bolivia have been given 12 million hectares of communal land under new legislation that recognizes the rights of indigenous communities to their original territories. The area under forest concessions declined from 15 million to 3 million hectares as a result of a new forestry law enacted in 1994; much of this land has gone to the indigenous territories (Pacheco and Mertens, 2004).

Forest landscape dynamics in the Central Highlands, Vietnam

Radical changes in Vietnam's economic policy since 1986 have included efforts to promote cash-crop cultivation in remote regions such as the Central Highlands. The government has promoted the transfer of people from overcrowded lowland areas to these new economic zones and spontaneous migration to these regions has vastly exceeded the rates that were envisioned in state planning. In 1921, the Central Highlands had a population of 98,000 inhabitants from 15 different indigenous ethnic groups, with almost no lowland Kinh inhabiting the region. This population structure remained more or less constant until 1975 and then increased to 1 million by 1976 and 4.2 million by 2002. Currently, local ethnic groups account for only 20 per cent of the region's population. The region's forests, which had been very extensive up until 1960, were lost at a rate of 30,400 hectares per year between 1976 and 1990. Although this rate declined in the subsequent five years, it still remained at 25,200 hectares per year between 1991 and 1995 (Tran Van Con, forthcoming). The highlands now have 57 per cent of their original forest cover.

At the same time, the state-imposed administration of natural resources has profoundly affected the self-control mechanism that was based on customary laws. In the customary societies of highland ethnic groups, community forests were clearly defined by village boundaries. These forests were managed by a self-rule mechanism based on highly effective village laws and regulations. New migrants arriving in the Central Highlands have affected existing land-tenure patterns. Land tenure is entirely customary based, as there are no officially recognized cadastral maps. As a result there are no mechanisms to allocate land to newcomers. At the same time, local communities have little interest in pursuing official recognition of land held as customary property. The demand by immigrants for both arable and residential land has led to illegal land trade, from both the state and customary perspective. Poor farmers often rent or sell part of their land to outsiders and then move deeper into the forest to occupy new land.

the model presented in Table 6.1 allows us to roughly gauge the position of a particular landscape within this process of land-use intensification and forest depletion. Many of the underlying forces of landscape dynamics are beyond the control of agencies in charge of forest or forest-landscape management. The following steps, however, allow forest-landscape managers to understand and take account of those dynamics over which they can exert some influence.

Step 1: Define adequate units and boundaries in the landscape of interest

- identify the landscape area whose dynamics you wish to understand; and
- identify the components of the landscape mosaic (for example, different categories of vegetation cover, land use or land tenure).

Step 2: Identify the relevant stakeholders

- identify the stakeholders in each of the components of the landscape.

Step 3: Identify the actions of relevant stakeholders and their impact on the forest landscape

- identify those actions that increase or decrease forest cover and those that improve or worsen forest condition; and
- identify those stakeholders that are having the largest (positive or negative) impacts on the landscape.

Step 4: Identify links

- explain the actions identified under the previous step to answer the question: Why are these actions taking place? This is the most comprehensive step in analysing the underlying forces of a particular forest landscape dynamic and involves the following actions:
    - Define the economic environment that leads to the stakeholder actions you are trying to explain. Are the stakeholders responding to certain economic incentives? Which resources are of economic value? Provide a characterization. Why do they have a value – for example, is it because of good markets for their products or because of low procurement costs?
    - Identify the policy environment. Are the stakeholders responding to policy incentives or opportunities? There may be legislation that enables these actions or there may be more specific circumstances such as tax breaks that the government is using to stimulate certain sectors (including forestry, agriculture or mining).
    - Identify the political position of the different stakeholders. In many instances stakeholders have certain political positions or connections with influential groups that provide them with opportunities or advantages over others.

Step 5: Analyse the results

- design a conceptual model of the stakeholders, their actions and how they link to explanatory factors; represent this graphically; and
- develop a written version of the conceptual model; this written version contains more detail and clarification, and can be used for reporting on the analysis and proposing recommendations for action.

A variety of methods can be used to complete these five steps. Techniques for steps 1 and 2 are described in Chapters 5 and 7 respectively. Step 3 can be completed through direct observations and consultation of existing reports prepared by government bodies, environmental NGOs or other agencies. Step 4, in many cases, can be completed through interviews with key informants – people who have specialist knowledge of, or long-term experience in, a particular issue or area. Existing reports are another source of information for this step and can include internal reports, published reports, academic papers, newspaper articles, editorial comments or dissemination material.

Step 5 involves constructing a conceptual model and a graphic representation of the analysis. One means of doing this (as shown in Figure 6.1) is to draw a central shape (a square or a circle) and divide it into different forest landscape components. Then link to each component, via arrows, a series of boxes with the names of key stakeholder groups that influence the component. Above the arrow insert a few words on the sort of influence that the stakeholder group has on the landscape component and the main motives

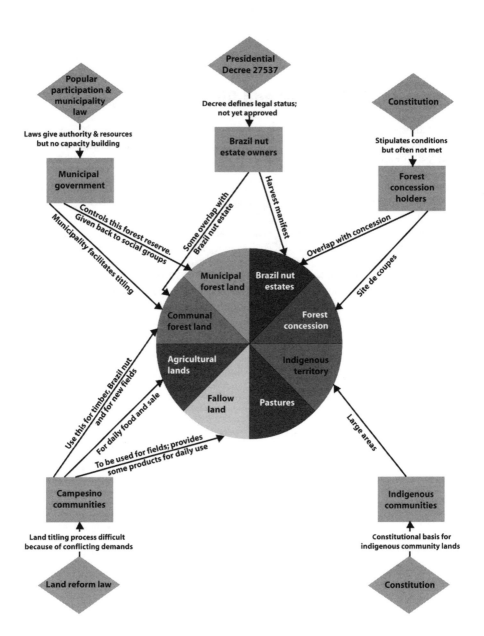

Figure 6.1 *Graphic representation of the analysis of forest landscape dynamics*

of the group. Additional shapes can then be added to provide information on other stakeholders or conditions that influence these key stakeholders. In this way the direct causes of landscape change are placed in close proximity to the landscape components and the indirect, underlying causes of change are represented as more peripheral influences.

References and further reading

Chokkalingam, U., Smith, J., de Jong, W. and Sabogal, C. (2001) 'A conceptual framework for the assessment of tropical secondary forest dynamics and sustainable development potential in Asia', *Journal of Tropical Forest Science*, vol 13, no 4, pp577–600

Pacheco, P. and Mertens, B. (2004) 'Land use change and agriculture development in Santa Cruz, Bolivia', *Bois et Forêt des Tropiques*, vol 280, no 2, pp29–40

Sheil, D., Puri, R. K., Basuki, I., van Heist, M., Wan, M., Liswanti, N., Rukmiyati, Sardjono, M. A., Samsoedin, I., Sidiyasa, K., Chrisandini, Permana, E., Angi, E., Gatzweiler, F. and Wijaya, A. (2003) *Exploring Biological Diversity, Environment and Local People's Perspectives in Forest Landscapes*, 2nd edition, Center for International Forestry Research, Ministry of Forestry and ITTO, Bogor, Indonesia

Tran Van Con, 'Migration and the social ecology of tropical forests in the Central Highlands of Vietnam', in W. de Jong and K. Abe, (eds) *Migration and the Social Ecology of Tropical Forests* (in preparation)

7

Applying a Stakeholder
Approach in FLR

Trikurnianti Kusumanto

This chapter looks at how FLR initiatives use a stakeholder approach to identify, understand and address the interests and concerns of key stakeholder groups. This kind of approach is important in FLR for two reasons. First, the success of FLR initiatives will depend on the willingness of stakeholder groups to cooperate with each other and with the FLR efforts. Second, since some stakeholders will be affected by the FLR activities, they need to be involved in decisions regarding the goods, services and processes of the landscape that are to be restored. Thus a stakeholder approach will help achieve the goal of equitable benefit-sharing among the key stakeholder groups. However, despite the importance of stakeholder approaches to FLR, caution is required when using them (Box 7.1).

This chapter sets out the four steps involved in a stakeholder approach:

1   understanding the context of stakeholder processes;
2   identifying the key stakeholders;
3   understanding stakeholder interests and interactions; and
4   managing multi-stakeholder processes.

Understanding the context of stakeholder processes

FLR practitioners need to understand the context in which they will work with stakeholders and be aware of why stakeholder involvement is critical to their work. The importance of stakeholder involvement stems from various aspects of the natural resource management context, including the following:

Box 7.1 Exercising caution in applying a stakeholder approach in FLR

Bear in mind that:

- it is not always possible to assign distinct identities to stakeholders, as they are often engaged in many overlapping roles and activities and these can change over time; and
- information about the interests of less-powerful stakeholders that is revealed and openly discussed as part of FLR activities can be misused by power-holders to further their own interests.

To address these issues:

- pay heed to possible consequences when using the approach;
- be prepared to adapt the approach to local circumstances, adjusting it whenever necessary through adaptive management; and
- consult the less-powerful stakeholders before any wider stakeholder meetings in order to learn about their particular concerns.

- natural resource management issues cut across social, economic and political spheres and involve many different stakeholder groups;
- natural resource management issues are often on a large scale (covering, for example, a watershed, province or nation). This means that some stakeholders may have to bear the costs (or enjoy the benefits) generated by the management actions of other stakeholders. For example, the excessive use of fertilizers by upstream farmers may pollute the soil cultivated by downstream villagers;[1] and
- use rights over resources can be unclear, conflictive or open to common-property resource problems. In such situations stakeholders may compete with each other for the available resources.

Box 7.2 shows how different stakeholder groups may have quite different views on what makes up a forest landscape.

Identifying the key stakeholders

A stakeholder, as defined here, is an individual, group of people or organization that can directly or indirectly affect the FLR initiative or be directly or indirectly affected by it. Key stakeholders need to be identified early on in an FLR initiative, as the information revealed may influence the activities and results of the restoration work. This identification will then need to be revised, reviewed

Box 7.2 Different landscapes for different stakeholders: A case from Bolivia

Since the enactment of Bolivia's new forestry law in 1994, indigenous and other rural communities have become a principal stakeholder group in the country's forest management (see also Box 6.1). Large areas of forest land have been designated solely for community use. These communities define the landscape in terms of the demands they place on it – for providing medicinal plants for local use, sustaining the local economy and securing the livelihoods of future generations. Other stakeholder groups define the landscape in different terms. For those with concession rights over forest lands, the landscape is defined by the financial benefits that the forest provides, while for an ecotourism operator it is defined by its biodiversity and cultural values.

and revisited at later points throughout the FLR initiative; stakeholders initially identified as key may later become less relevant and new groups may become apparent only during later stages of the restoration. For this reason, stakeholder identification and verification should be viewed as a continual and ongoing process that is undertaken alongside the actual fieldwork.

In working with the different stakeholders, FLR practitioners need to accommodate their different definitions of the relevant landscape. Appreciation of these definitions (which may be quite different from their own) is important in creating space for negotiating the objectives and outcomes of the FLR work. At the same time, however, practitioners need to keep in mind the overall aims of the FLR – that is, to restore ecological integrity and enhance human well-being.

Box 7.3 lists some of the questions that can be asked when identifying stakeholder groups.

Box 7.3 Useful questions to guide stakeholder identification

- Who is likely to be affected by the FLR initiative, either positively or negatively?
- Who will make the FLR initiative more effective by participating (or who will make it less effective by not participating)?
- Who may oppose the FLR initiative? What can be done to encourage such stakeholders to cooperate?
- Who will be able to contribute to the FLR initiative with knowledge, skills and other resources?

Box 7.4 Common approaches to identifying stakeholders

Identification by the stakeholders themselves: staff of FLR initiatives disseminate information through the local media or during field visits and invite stakeholders to come to meetings.

- *Risks*: those with less access to media may miss the information. The less-educated and less well-off may be reluctant to come forward and those sceptical about the initiative may not want to participate in the meetings.

Identification by other stakeholders: stakeholders identified at an early stage can then become sources of information on other stakeholders. This approach can be helpful in identifying those people whom stakeholders consider to be representatives of their or other groups or those considered important for other reasons.

- *Risks*: stakeholders may, when consulted about other stakeholders, be selective in whom they propose, based on their personal preferences.

Identification by knowledgeable individuals or groups: knowledgeable individuals (key informants) or groups can help identify stakeholders. These individuals or groups may include village elders, women, forestry agency staff or neighbouring communities.

- *Risks*: less 'visible' stakeholders may be under-represented.

Identification by field-based staff of the FLR initiative: staff who have worked and lived in the area for some time may have valuable knowledge for the identification of stakeholders.

- *Risks*: staff may select the same individuals or groups with whom they have worked before. Women may be under-represented.

Identification based on demography: social groups are identified based on their demographic characteristics, such as age, occupation and gender.

- *Risks*: when using many characteristics, the number of stakeholders identified may become too high, making the management of the FLR implementation stage difficult.

Identification based on written records: forest agencies, local NGOs and training institutions often keep records that can be useful in identifying stakeholders. For example, these records might provide baseline data, population data, data on conflicting groups or lists of licence-holders.

- *Risks*: the written information that is available may not always be accurate, complete or up-to-date. Also, bias may have crept into the written reports.

A wide variety of handbooks and manuals is available to practitioners to help answer these questions.[2] Some of the most commonly used approaches are presented in Box 7.4, along with the risks or shortcomings of each. In order to minimize the risks and ensure identification of all the relevant stakeholder groups, it is best to use a combination of the different approaches.

Understanding stakeholder interests and interactions

Having identified the relevant stakeholders for the FLR initiative, practitioners then need to learn about the interests of, and interactions between, the different stakeholder groups. Some information on this will probably have been gathered during the stakeholder identification process; this can serve as the basis for further investigation. The key objective of this stage is to ascertain how stakeholders see their current and potential role in resource management within the forest landscape. Box 7.5 lists some questions that practitioners can use at this stage; the set of questions is illustrative rather than exhaustive and each FLR initiative will require its own specific list to be drawn up.

Box 7.5 Questions for investigating stakeholder interests and interactions

- How do stakeholders use and manage resources in the forest landscape? What goods and services do they get from these resources? What goods and services do the stakeholders provide? Are there restrictions over the use of the resources? What are the stakeholders' official and informal rights over these resources?
- What are the views of different stakeholders on the role of other individuals or groups in the use and management of resources? Do they use and manage the same resources as these other stakeholders? If so, how do they interact with these other stakeholders?
- How do stakeholders make decisions on the use and management of resources? What criteria do they consider when choosing a particular option?
- How far do stakeholders think their decisions will reach? What factors lie within their control and what beyond it?

Again, there are various tools available for collecting this kind of information. Some of the most commonly used tools include various participatory rural appraisal techniques, focus-group discussions and semi-structured interviews.[3] These should be complemented by different methods, including direct observation of stakeholder actions and behaviours, to cross-check the information obtained.

When exploring stakeholder interactions, practitioners should look out for any situations of conflict or trade-off (see Box 7.6). Understanding the conflicts between stakeholder groups is a necessary first step of any conflict management strategy. Similarly, understanding the trade-offs involved in choosing between mutually exclusive objectives will help practitioners to encourage stakeholders to see the value of FLR and to better manage the process.[4]

Box 7.6 Conflicts and trade-offs

A *conflict* is a situation of disagreement between two or more different stakeholders or stakeholder groups. In some cases, there may also be internal conflicts within stakeholder groups. Conflicts are normal wherever human-beings interact and do not always involve violence. Conflicts can be managed constructively (see Chapter 13).

A *trade-off* is a situation where a balance needs to be reached when choosing between two desirable but incompatible objectives or outcomes. Trade-off situations are the rule rather than the exception in natural resource management. The successful implementation of FLR requires that trade-offs are made explicit and joint solutions sought.

One way to learn more about conflict situations is to discuss a past conflict in order to find out:

- who was involved;
- what gave rise to the conflict;
- how it was resolved or managed; and
- if a conflict remained unresolved, why this was the case.

In the case of serious disagreements, it is usually best to approach stakeholders separately to address these questions. If, however, group meetings cannot be avoided, intermediaries who are respected and considered impartial by all parties can be asked to moderate these.

Once the information on stakeholder interactions has been collected, it needs to be organized in a way that facilitates further analysis and discussion. One useful tool for doing this is a matrix, such as the one shown in Figure 7.1, which summarizes the existence and level of conflict between different stakeholders over a particular landscape resource. The matrix also highlights any internal conflicts within stakeholder groups (as can be seen, in this hypothetical case, within government and within communities living in or close to the forest).

Another tool for assessing stakeholder interests is the '4Rs Framework', which sets out stakeholder rights, responsibilities, returns and relationships.[5]

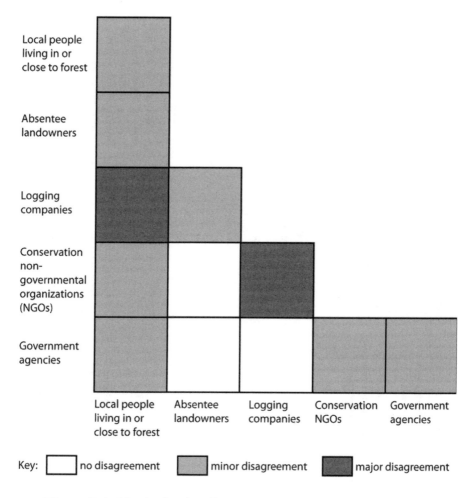

Figure 7.1 *Matrix showing disagreements between stakeholders over a landscape resource (hypothetical example)*

Source: adapted from Grimble et al (1995)

Table 7.1 presents an example of such a framework from Indonesia, where this tool was used in preparation for an action learning process involving different stakeholder groups within the context of a collaborative forest management project.[6] The matrix makes explicit several imbalances in stakeholder roles and responsibilities. For example, those with the most stake in the forest (that is, the original inhabitants) have limited legal responsibilities related to forest management. On the other hand, while the government has the responsibility to manage and protect the forest, they lack the means to do this effectively. In principle, responsibilities (and therefore rights) should be transferred to those who have more stake in the forest and arrangements created for effective

Table 7.1 *The 4Rs framework: An example from Jambi, Sumatra*

Stakeholders	Rights	Responsibilities	Returns	Relationships
Nomadic group (Orang Rimba)	• Customary rights (for which official recognition should be sought) • Limited formal rights, particularly because the group has no administrative 'home'	• Traditional management and protection of natural resources • No formal, legal responsibilities related to natural resources	• Non-timber forest products (NTFPs), crops and other forest goods • Environmental services, homesteads • Social security from patron–client relationship with some villagers	• Customary rights over land and forest resources not recognized by the state • Weak relationship with villagers • Weak relationship with public bodies • Patron–client relationship with some villagers
Original inhabitants	• Customary rights (for which official recognition should be sought) • Limited formal rights	• Traditional management and protection of natural resources • No formal, legal responsibilities related to natural resources • Pay taxes	• Timber and NTFPs, crops, income and other forest goods • Environmental services • Benefits from land (including grazing)	• Customary rights over land and forest resources not recognized by the state • Poor relationship with government because traditional shifting cultivation is officially not recognized and because customary land has been allocated to settlers • Poor relationship with settlers because the latter were officially allowed to 'occupy' customary lands
Settlers	• Formal rights over registered land-holdings under resettlement programmes (rights of inheritance and land transaction)	• Develop agricultural landholdings under resettlement programme • No formal, legal responsibilities related to forest resources	• Annual crops from dry swiddens • Crops and perennial products from registered landholdings under resettlement programme	• Poor relationship with original inhabitants because of 'occupation' of customary lands • Little commitment to resource management

Stakeholder	Rights	Responsibilities	Interests/Benefits	Relationships
				and protection other than on their own agricultural holdings
Sawmill owners, small-scale timber investors, middlemen, loggers	• Illegal sawmill owners hold no official rights • Official licence-holders hold official permits	• Respect customary authority of original inhabitants over land and tree resources • Pay taxes • No formal, legal responsibilities if illegal • Pay levies in the case of licence-holders	• Income from the sale of products, providing services or wage labour	• Working and commercial relationships with original inhabitants and some settlers • Poor relationship with government in the case of illegal sawmill owners • Official relationship with government in the case of licence-holders
Government logging company	• Logging rights	• Community development • Job creation • Sustainable natural resource management practices	• Financial benefits • Financial objectives met • Income • Homes for staff	• Poor relationship with original inhabitants • Official relationship with local government
District forestry agency	• Rights to give permits regarding forest products (including timber) • Rights to arrest illegal users • Rights to propose resource management procedures	• Implement government forestry policies, programmes, and management plans • Arrest illegal users • Control implementation of management plans	• Policy and programme objectives met • Prestige (respect/fear) • Recognition of authority • Financial benefits	• Limited relationship with original inhabitants, mostly during incidental monitoring visits
NGO implementing the integrated conservation and development project	• Rights to develop and manage park and buffer-zone implementation plans • No legal rights to forest	• Develop and implement park and buffer zone management plans • Coordinate with national park agency for project implementation	• Project objectives met • Jobs	• Relationship with original inhabitants and settlers limited to project activities • Official relationship with local government

relationships between stakeholders. The role of an FLR practitioner is to assist stakeholders in negotiating a more balanced set of '4Rs'.

This stage of an FLR initiative requires a considerable level of communication between FLR field staff and a variety of stakeholder groups to gather the necessary information. These interactions with stakeholders should be used as an opportunity to build trust with local groups, and this is also an appropriate point at which field staff can begin to systematically encourage communication and collaboration between the different stakeholders.

Managing multi-stakeholder processes

As discussed in Chapter 4, FLR is implemented using an adaptive management approach that involves an action learning process whereby stakeholders collaboratively, systematically and deliberately plan, implement and evaluate the restoration activities. Through this process of learning the stakeholders build experience in order to act collaboratively as a group. The role of FLR field staff here is to manage this process by facilitating collaboration between stakeholders.

To enable practitioners to take on this facilitation role, FLR initiatives need to develop appropriate arrangements for the action learning activities. These arrangements need not be set up especially for the FLR initiative – arrangements may already exist, such as community-wide meetings, encounters among neighbouring communities or government consultation meetings involving local groups and other stakeholders.

Whether newly established or existing, these arrangements should meet the following requirements if they are to support multi-stakeholder processes effectively:

- they need to ensure that ***all stakeholders*** are involved in the action learning, not just their representatives. All stakeholder representatives who are engaged in the FLR activities must be elected by group members rather than appointed and they must report back to their groups and consult with them before any binding decisions are made;
- they need to support the stakeholders in building ***joint experience***. This means that funds and other resources should be available for stakeholders to develop real experience;
- they must accommodate the different needs of the stakeholders so that ***ownership*** of processes is ensured; and
- they need to support ***fair communication*** among stakeholders by organizing activities that create a fair chance for all stakeholders to express their views, understand the views of others and be understood by others.

Once these arrangements have been established, FLR practitioners can begin the actual facilitation work. Several publications provide guidance on this topic;[7] here we look at two important aspects of the facilitation process: joint decision-making and conflict management.

Box 7.7 Effective facilitation of joint decision-making

Process management or facilitation involves the enhancement of learning and can only be effective if process managers deliberately put learning at the heart of activities. For an effective enhancement of learning, process managers or facilitators should:

- focus on the *core values* of joint decision-making. These values are a shared responsibility for the consequences of decisions; the inclusiveness of decisions; a mutual appreciation of one another's views; and active participation by all stakeholders. Joint decision-making means that facilitators do not make decisions themselves but guide the process by which the different stakeholders collectively arrive at decisions;
- have the appropriate *attitude*. This implies that process facilitators should have a sense of fairness so that the stakeholders consider the facilitated processes fair. Facilitators should also be empathetic and good listeners. Having the appropriate attitude is more important than any facilitation or learning tool. Facilitators' ability to adopt the right attitude can improve as they gain more experience with multi-stakeholder processes;
- provide the right *conditions* for stakeholders to learn new ways of joint decision-making. There are three important conditions here. First, stakeholders need to feel encouraged to propose new, creative ideas, however absurd these may seem. The more creative the group and the more alternative the decisions proposed, the more likely it is that an innovative decision will be taken. Second, stakeholders should be encouraged to take time to think and to reflect critically on their assumptions and old ways of thinking. Third, the facilitation should aim to build constructive relations between the stakeholders; and
- be equipped with effective *tools* to facilitate group processes. Effective facilitation tools are those that encourage joint learning and may include participatory mapping, focus group discussion, brainstorming, community meetings, scenarios, role-plays and computer-based simulation modelling.

To facilitate joint decision-making effectively, practitioners need to:

- focus on the *core values* of joint decision-making;
- have the appropriate *attitude*;
- provide the right *conditions* for stakeholders to learn new ways of making decisions; and
- be equipped with the *tools* for the facilitation work.[8]

These points are expanded in Box 7.7 and one of the tools, brainstorming, is explained in Box 7.8.

Box 7.8 Brainstorming as a facilitation tool for joint decision-making[9]

A helpful tool for facilitating the process of joint decision-making is brainstorming, which can be used within an FLR context to examine the causes of prevalent resource problems in the landscape, look for potential solutions to declining resource availability, explore marketing options for NTFPs or make sense of the reasons why certain groups oppose collaborative solutions.

One principle of the brainstorming tool is that 'anything goes', in other words that the listing of perspectives and ideas is made without any censoring or debating. Brainstorming should also be accompanied by techniques to conclude the brainstorming session, such as the clustering or prioritizing of decision alternatives.

The role of process facilitators is in guiding group processes to explore and synthesize decision alternatives to arrive at a decision that works for everyone.

The topic of conflict management is widely covered in training events and materials.[10] Here are some basic tips for practitioners working in situations of stakeholder conflict:

- Make assumptions explicit so that they will not hinder communication.
- Make one of the facilitation goals the development of constructive relationships.
- Be clear to all parties about the outcome of group processes within the negotiations. Will the group processes stop at knowledge generation or will they result in decisions being taken?
- Ensure that negotiation between parties reaches out to all members of stakeholder groups and does not only involve their representatives.
- Be constantly aware that conflicts often involve ordinary members of stakeholder groups, not just their representatives or leaders.
- Be alert when conflicts intensify in order to bring in conflict mediators at the right time.

Notes

1   This situation is sometimes explained by saying that 'externalities' occur.
2   See, for example, Grimble et al (1995), Higman et al (1999), Colfer et al (1999) and Richards et al (2003).
3   See Pretty et al (1995) for a useful and practical manual that shares participatory methods of data collection and analysis.

4   See Chapter 3 for more on how the double-filter concept can help address trade-off issues in building support for FLR.
5   See Dubois (1998) for further guidance in developing such a framework.
6   See Kusumanto (2001).
7   See references listed in Footnote 10 and also Braakman and Edwards (2002), Hartanto et al (2003) and Kusumanto et al (2005).
8   See, for example, Wollenberg et al (2000) and Nemarundwe et al (2003) for guidance on scenarios; see also www.cifor.cgiar.org/ACM for the computer software package Co-Learn, which supports the joint management of natural resources by helping people to enjoy learning processes in groups.
9   Adapted from Braakman and Edwards (2002).
10   See, for example, Means et al (2002) for conflict management training guidance and www.recoftc.org or www.iac.wur.nl for training concerning stakeholder conflict management.

References and further reading

Braakman, L. and Edwards, K. (2002) *The Art of Building Facilitation Capacities*, Regional Community Forestry Training Center for Asia and the Pacific, Bangkok, Thailand

Colfer, C., Prabhu, R., Günter, M., McDougall, C., Porro, N. and Porro, R. (1999) *Who Counts Most? Assessing Human Well-Being in Sustainable Forest Management*, The Criteria & Indicators Toolbox Series No 8, CIFOR, Bogor, Indonesia

Dubois, O. (1998) *Capacity to Manage Role Changes in Forestry: Introducing the '4Rs' Framework*, International Institute for Environment and Development (IIED), London, UK

Grimble, R., Chan, M., Aglionby, J. and Quan, J. (1995) *Trees and Trade-offs: A Stakeholder Approach to Natural Resource Management*, IIED, London, UK

8

Identifying Site-level Options

David Lamb

This chapter explores how biophysical, socio-economic and silvicultural factors can determine the feasibility and suitability of different restoration options for particular sites in the landscape.[1] The variety of ecological conditions and diversity of stakeholder views mean it may not be possible to restore forest to all sites in a landscape. Furthermore, different stakeholders will have different objectives when they carry out reforestation. However, by strategically targeting areas for various kinds of reforestation, these interventions will collectively improve key ecological processes (hydrological functions, nutrient cycling, etc.), restore biodiversity and thereby improve livelihoods across the landscape. Thus a landscape mosaic after restoration might include land uses such as:

- areas managed to maximize production (of, for example, agricultural crops or pulpwood plantations);
- areas of existing forest managed to maintain existing levels of biodiversity (for example, natural forest devoted to nature protection or well-managed natural forest used for timber production or the harvesting of non-timber forest products (NTFPs)); and
- reforested areas managed to generate both commercial outcomes and restore some, but not necessarily all, the original biodiversity (for example, long-rotation sawlog plantations containing high-value native tree species).

The range of site-level options is described in Chapters 9, 10 and 11.

Biophysical factors affecting restoration choices

Deforestation and degradation can cause large changes in a variety of biophysical factors and these will constrain the types of restoration that might be carried out. The key biophysical variables include the degree of deforestation and forest fragmentation that has occurred, the levels of soil fertility at the deforested sites, and the topography and microclimates across the landscape. These factors might vary over time because of natural or human-induced disturbances.[2] The influence of these biophysical factors on restoration options is outlined in Table 8.1.

Table 8.1 *Biophysical factors that can influence restoration choices*

Biophysical factor/ feature	Influence on restoration choices
Area of residual, largely undisturbed natural forest that remains	Extent and distribution determines how many of the original goods (including timber and NTFPs) and services (such as watershed protection or biodiversity) are still being supplied. This in turn determines how attractive any kind of restoration at the other more disturbed and degraded areas may be to stakeholders (restoration is likely to be more attractive when the remaining area of natural forest is small).
Area of secondary or regrowth forest	Many of these forests are still able to supply goods (especially to local communities) and services. Areas on hills may be especially important for watershed protection.
Quality of secondary forest	The rate of recovery and the nature of the goods supplied by these forests depend on species composition, stand structure and regenerative capacity. Some will recover without further intervention if they are simply protected, but the rate of recovery and the types of goods supplied can often be accelerated by silvicultural interventions.
Quality of agricultural land	Restoration is likely to be most attractive if there are large areas of low-productivity land because the opportunity costs of reforestation will be lower (reduction in agricultural production that occurs when these lands are reforested will be less).
Amount of unused and degraded land	The larger the area of degraded land, the greater the ecological and social benefits that are likely to result from restoration. If the land is unused, there is unlikely to be any loss of agricultural production when reforestation takes place[3].
Environmental priority areas	These are areas with significant environmental problems, such as uplands with eroding slopes or polluted areas affecting other land users in the landscape. They may also be areas with particular weeds or pests. These are likely to be priority areas for restoration.

Table 8.1 *continued*

Biophysical factor/ feature	Influence on restoration choices
Areas difficult to reforest	These difficult areas might be sites with shallow, especially infertile or polluted soils, swampy sites, or landslip-prone areas. Restoration of these sites might require unusual and expensive approaches. These areas may be too costly to treat and, in some special cases, it may even be preferable to leave them untouched and to treat larger areas at less difficult sites.
Areas of biological significance	These areas may be sites with unusually high levels of biodiversity or sites containing habitats of important plant or wildlife species. If the areas are small or threatened by new disturbances, they might be protected by creating forested buffer zones around each site.
Accessibility of sites	Degraded sites or regrowth forests that are difficult to reach will be expensive to restore. In many situations it may be too difficult to do anything about such sites.
Climatic seasonality	Other things being equal, it is usually much easier to carry out restoration in a non-seasonal, wet tropical landscape than in a monsoonal forest landscape that is subject to a prolonged dry period. Little can be done about any climatic constraints other than to work with tree species adapted to such climates.

The main conclusions to be drawn from this are:

- prevention of further degradation is usually much cheaper than restoration, so every care should be taken to protect remaining natural forests;
- secondary forests are also crucial for FLR since the cost-effectiveness of restoring these is likely to be much higher than that of most other forms of restoration;
- restoration should be carried out in areas where the opportunity costs are low (where, for example, the fertility is low and the land is less attractive for food production) and where the functional benefits will be high – for instance, in erosion-prone or highly degraded areas;
- there may be priority areas deserving early attention (such as point sources of erosion on river banks, unstable hill slopes or areas of conservation value that are at risk);
- it is often useful to target areas around remnants of existing forest for restoration, since these reforested areas can then act as buffer zones and help prevent further degradation of the remnants; and

- the way in which an intervention is actually carried out – and the choice between spending more resources on restoring secondary forest or replanting completely deforested sites – will depend on locally determined priorities.

Restoration is often a new land-use activity for many landowners and care needs to be taken to demonstrate its value. It can be useful to tackle less difficult sites first. These early sites will then act as demonstration areas,[4] and success breeds success. It is also useful to have these demonstration sites located in different parts of the landscape, to reach a wide range of stakeholders. In selecting areas of forest landscape to restore, it is important to consider scale requirements for the objectives of the restoration work, as outlined in Box 8.1.

Box 8.1 The importance of scale

Different processes operate at different scales. A small area of forest or plantation may be sufficient to prevent erosion from a localized source. A narrow strip of trees might be effective as a windbreak or to help stabilize a hillslope. But an isolated patch of forest of the same size is unlikely to be useful in conserving biological diversity. Size does matter, and larger areas are usually better than small ones, especially in highly fragmented landscapes.

Socio-economic factors affecting restoration choices

Most forest degradation has a socio-economic cause and there is little point in trying to restore any degraded forest land unless the past and present socio-economic causes of the degradation have been investigated and understood. For example, it may be quite possible to undertake restoration if the cause of damage was simply poorly supervised logging and if few people currently live on the site. But it may be much more difficult if degradation was caused by increasing populations of recent migrants clearing forest in search of new agricultural land. This is because many small land-users may be benefiting from the present 'degraded' landscape and may be reluctant to change their land-use practices, even if the wider community is being disadvantaged by their activities.

A summary of some of the key socio-economic factors that may determine the attractiveness of restoration options to local communities is shown in Table 8.2. All of these factors can affect whether individual land users believe restoration is likely to benefit themselves and their families. Farmers will judge how to maximize the financial benefits arising from different alternatives (and how to minimize the risks involved). The most attractive options are likely to be those where benefits are quick to appear.

Table 8.2 *Socio-economic and cultural factors that can influence the attractiveness of restoration to communities living in degraded forest landscapes*

Socio-economic/ cultural factor	Significance
Availability of agricultural land	Reforestation of any kind will be difficult if there is a shortage of land for food production. In these circumstances it may even be difficult to protect any large residual forest areas.
Land tenure and patterns of land use	Land users are only likely to participate in restoration if they or their families will benefit. This is unlikely if they have no tenure. Restoration that results in reduced access to land that is currently available will be unattractive unless some form of compensation is available.
Degree of dependency on traditional forest products	Restoration is more likely if the supply of valued forest goods (such as medicinal plants) from natural forests is declining and there are no alternative supplies.
Knowledge of markets for timber and other forest products and services	Restoration is easier if it there is a known market (and especially an improving market) for forest goods and/or services, particularly if further supplies from natural forests are unavailable.
Existing plantations	These provide a benchmark (i.e. species used, growth rates achieved, markets supplied) for use in planning rehabilitation. Such demonstration areas can be especially valuable where farm forestry has not been a traditional land use.
Timing of financial benefit	Interventions that produce early cash flows (from, for example, agricultural cash crops) are more attractive than those where financial benefits are delayed (such as in the case of sawlog plantations). There may be scope for blending the two (for example, by growing NTFPs in plantation understoreys).
Risk involved	Low-cost restoration (such as fostering secondary forest recovery) is likely to be less risky and more attractive than higher-cost methods (such as plantation establishment). Similarly, fast-growing species are usually more attractive than slow-growing species, particularly when the harvesting period is distant. Financial incentives or subsidies can sometimes reduce this problem.

Table 8.2 *continued*

Socio-economic/ cultural factor	Significance
Access to finance	Restoration is very expensive. Particularly in the initial stages, low-cost finance, subsidies or incentive payments may be needed to achieve significant change. Payments for the supply of ecological services may be especially attractive. These can be used to foster landscape-level objectives and priorities.
Attitudes of neighbours	Uncooperative neighbours who illegally remove forest products or allow fires, grazing stock or weeds to cross into regenerating forest can easily disrupt rehabilitation and restoration. Ways should be found to ensure their collaboration.

Cultural attitudes are also important. Communities differ in the extent to which they see tree-growing as a traditional or useful practice. Many traditional communities routinely cultivate certain trees for a variety of economic, medical, social or cultural purposes. In contrast, farmers who are more concerned with livestock production, or migrants who have only recently moved into an area, may be unfamiliar with local forests or tree species and less inclined to engage in reforestation.

The scale of tree-planting or farm forestry carried out in FLR is likely to be greater than many landholders may have been involved in before. In some cases, farmers can be quickly convinced of the benefits of reforestation, particularly if financial assistance is available or there are clear market opportunities. Knowledge is also important and some farmers will quickly learn from demonstrations or be willing to try new tree seedlings offered to them by extension officers. On the other hand, other farmers may be more sceptical of government officials and will prefer to learn about the techniques and benefits of tree-planting from their neighbours.

Tree-growing can be a commercially risky business since it is difficult to be certain about future markets. Under these circumstances risks can be reduced by using fast-growing species. But this approach is not without its own risks. Box 8.2 illustrates one case where tree-planting with a small number of fast-growing species has provided farmers with relatively little financial benefit. An alternative strategy might be to concentrate on slower-growing but higher-value species and to use a variety of these to spread risk.

The main lessons emerging are:

- most landscapes will have many stakeholders with different priorities and different socio-economic characteristics;

Box 8.2 Single options can be risky: The case of Vietnam

Vietnam has embarked on a large-scale reforestation programme. Since many sites are degraded and infertile, fast-growing exotic species such as eucalypts have been used widely. Although these species have restored forest cover successfully, their financial benefits to farmers have sometimes been disappointing. As plantations have matured, large volumes of eucalypt timber have come on the market and prices paid to small growers have declined. By reducing the heterogeneity of the landscape, the restoration programme has increased the risks not only of future market problems but also of possible outbreaks of pests and diseases. As a general principle, FLR should increase, not reduce, the heterogeneity of the landscape.

- under these circumstances legal constraints may also be needed to limit further degradation;
- the attitudes of stakeholders are determined largely by self-interest. Not surprisingly, poor farmers give priority to food production above all else. Other stakeholders may be attracted by the perceived benefits of a future supply of other forest-based goods or services;
- many traditional communities are dependent on goods and services derived from forests and will be interested in forest restoration because it will enhance the supply of these;
- tree-growing is a long-term enterprise and ways must be found to make the long-term benefits as appealing as the short-term benefits;
- FLR practitioners will need to determine what incentives or compensation (financial or otherwise) might be needed to make FLR more attractive to stakeholders and especially to local land users; and
- the difficulty of predicting future markets may be at least partly overcome by including a variety of species and giving preference to species likely to have higher market values.

Ecological factors affecting restoration choices

In addition to these biophysical and socio-economic constraints there is a third group of factors that affect restoration choices. A range of ecological factors operate at the site level and dictate what types of silvicultural approaches might be possible in order to achieve restoration at a given site. Some of these limiting factors are summarized in Table 8.3.

Since different parts of the landscape will be affected by some or all of these ecological and silvicultural factors, FLR practitioners will usually need to

Table 8.3 *Ecological factors that can influence restoration choices*

Limiting factors	Significance for replanting or successional development
Existing tree cover	The amount of canopy tree cover will determine whether it might be preferable to clear the site and replant or whether it is feasible to rely on natural regeneration.
Soil fertility	Soil fertility is initially a consequence of the underlying geology. But if topsoil erosion has occurred, many of the original species may now be unable to grow unless nutrient deficiencies are addressed. Tolerant pioneer species or even exotic trees will need to be identified for these sites. Such species can facilitate the later introduction of more preferred species. Fertilizers can help overcome deficiencies but are expensive to use over large areas.
Fire regime	Fires are more frequent in seasonal climates with long dry seasons; they are also often more common in degraded landscapes, perhaps because the original forests have been replaced by grasslands. Ways must be found to reduce fire frequency, at least until new forests become well-established. It may be useful to create buffer zones around the main restoration area using more fire-tolerant species.
Seed-dispersing agents	Seeds of many recolonizing species are dispersed by birds or bats. Not all these animals will move over deforested areas and the rate of dispersal diminishes when these sites are more distant from natural-forest remnants. This means isolated areas are unlikely to benefit from successional development.
Weeds	Many restoration projects fail because weeds are not controlled, although weeds usually become less of a problem after canopy closure. Grasslands are particularly difficult sites for woody plants to colonize.
Pests	Some animal species, especially herbivores, can destroy young seedlings; ultimately, fencing may be needed.

conduct a survey or site classification to determine which factors are present in each area. Box 8.3 provides an example of how poor soil fertility can determine silvicultural choices.

Box 8.3 Soil fertility constrains rehabilitation choices

Tin mining has been carried out in several areas of Malaysia and Thailand. Most of these sites have very sandy soils. Following removal of the original rainforest, dredge mining was used to extract the tin. This destroyed the topsoil and dramatically changed the soil conditions and fertility. The loss of topsoil meant that most of the nutrients, seed stores and mycorrhizae at the site were lost, which in turn has meant that none of the original tree species formerly growing at these sites can now reoccupy them. The absence of organic matter also means that any fertilizers applied to overcome the fertility problems are more easily lost through leaching. Trials are now under way to reforest these areas using a variety of species including exotic, nitrogen-fixing trees and shrubs. Only a few tree species may be able to tolerate and grow on these impoverished sites until organic matter has been built up in the topsoil once more.

Some form of intervention is usually needed to overcome the constraints imposed by these ecological factors. Table 8.4 outlines how one such factor – tree cover – determines the choice of intervention in a given site. Often the most important initial decision is whether to retain existing vegetation and rely on natural recovery processes or clear the site and replant. This will always depend on the amount and condition of any residual forest and other local circumstances. If there is sufficient residual forest remaining then natural successional processes will allow recovery to occur. If planting must be carried out, the key question is whether the desired species can tolerate the site conditions (the current levels of soil fertility, for example). If the desired species can be planted, there are various types of plantations that might be used depending on market needs and the balance to be struck between the need to improve production and the need to restore biodiversity. When the current site conditions are unsuitable for the desired species, an intermediate step might be needed involving nurse trees or species able to facilitate the subsequent establishment of the preferred species. Further details of rehabilitation and restoration methodologies are given in Chapters 9, 10 and 11.

Scenarios of different site-level options

The following scenarios illustrate how ecological and silvicultural factors might influence FLR choices.

Table 8.4 *Residual tree cover as a determinant of restoration options*

	Level of forest cover remaining		
	Some residual forest present	**No residual forest present at the site**	
Possible restoration options	*Option 1:* *Rely primarily on natural successional processes* • Protect forest and allow natural recovery to occur • Protect forest and manage trees to favour particular species (e.g. by tending or thinning) • Protect forest and enrich with commercially desirable species	*Option 2:* *Establish plantations using the preferred species* • Use mono-cultures of species able to tolerate site conditions (preferably native species) • Use mono-cultures but plant different species in different parts of the landscape according to site conditions • Use mono-cultures and underplant with agricultural or other NTFP crops • Establish multi-species tree plantations	*Option 3:* *Use nurse crops or species able to facilitate the establishment of more preferred species* • Use tree species that can tolerate existing site conditions and exclude weeds, provide shelter or improve soil fertility, and allow the subsequent establishment of the preferred species

Scenario 1 A landscape with extensive forest cover remaining, although much of it has been heavily logged

Condition: the landscape has a large area of forest remaining. The lowland areas are mainly secondary or regrowth forest while the uplands are still occupied by undisturbed natural forest. Agriculture is practised on only a small area of relatively flat land in the lowlands.

Suggested approach: where possible, protect the remaining forest areas from further major disturbances and rely on natural regeneration to overcome past degradation. Enrich the secondary forest where commercial considerations suggest this will be useful. If forest clearing is necessary for agriculture, prioritize those areas that are of least importance for conservation and strive to achieve or maintain good connectivity between forested areas.

Comment: in this case there is probably no need to invest too heavily in restoration because most of the original biodiversity is still present across the landscape. Natural successional processes will lead to recovery over time since the main ecosystem processes are essentially intact.

Scenario 2 *A more heavily degraded landscape*

Condition: only a few small fragments of natural forest remain and even steep hillslopes are now without tree cover. Most vegetation now present is grassland or shrubs and erosion is widespread. Productive agriculture is possible on flatter land in the valleys but only temporary cropping is carried out on steeper land because of erosion problems.

Suggested approach: exclude further agriculture on hilly land and carry out reforestation on steeper slopes to control erosion (since there is insufficient woody regrowth to rely on rapid natural regeneration). Use whatever species can tolerate these soils, including exotic species if necessary, but give priority to higher-value species if the plantations are to be harvested eventually and long rotations are being used. Harvesting for timber on such slopes should be controlled carefully to maximize watershed protection and may be inappropriate in some situations.

Comment: the key objective in these circumstances is to restore ecological services (such as watershed protection and biodiversity conservation) rather than to maximize timber production. The opportunity costs of halting agricultural production are low, since this low-yield, short-term agriculture was probably contributing little to the local farming community. Increased forest cover on the steeper slopes will help lead to more sustainable agricultural practices in the lowlands.

Scenario 3 *A productive agricultural landscape with many small forest remnants*

Condition: the landscape has been cleared extensively for agriculture. While farming is currently productive, the level of biodiversity present has been greatly reduced and the sustainability of agriculture may therefore be at risk.

Suggested approach: encourage farm forestry using a variety of high-value timber trees on under-utilized land, giving priority to areas where erosion is present (e.g. steep slopes or riverine areas) and where plantations could form linkages between existing forest remnants.

Comment: functional benefits and biodiversity values will be improved if reforestation can be carried out at certain key locations across the landscape. This may be difficult to achieve where there are many landowners and may be especially difficult when landholdings are small. However, carefully located plantations involving a variety of higher-value species are likely to enhance both ecological and economic resilience, reduce risk, and lead to more sustainable systems of agriculture.

Notes

1   Chapter 13 takes this further by examining how compromises are managed to optimize outcomes across the landscape.
2   See Chapter 6 for more discussion on disturbance and dynamics in forest landscapes.
3   See, however, the discussion on supposedly 'unused' land in Chapter 5.
4   See Chapter 3 for a discussion on building support for FLR initiatives.

References and further reading

ITTO (2002) *ITTO Guidelines for the Restoration, Management and Rehabilitation of Degraded and Secondary Tropical Forests*, ITTO Policy Development Series No 13, ITTO, Yokohama, Japan

Lamb, D. and Gilmour, D. (2003) *Rehabilitation and Restoration of Degraded Forests*, IUCN, Cambridge, UK

Regional Community Forestry Training Center for Asia and the Pacific website: www.recoftc.org

9

Site-level Restoration Strategies for Degraded Primary Forest

Cesar Sabogal

This chapter, together with Chapters 10 and 11, sets out site-level strategies and their associated silvicultural techniques for restoring degraded primary forest, managing secondary forest and rehabilitating degraded forest land within the context of an FLR programme. These chapters are intended to provide an overview of the different strategies and methods, along with basic practical information on their use and guidance as to which methods are most appropriate in different situations. References are provided to point readers to more detailed technical guidelines.

Table 9.1 summarizes the restoration objectives and methods most suited to different types of degraded forest and other lands.

This chapter outlines the main strategies and silvicultural options available for the restoration of degraded primary forest[1], with particular emphasis on tropical rainforests. It provides some basic advice on how to choose the most appropriate methods and suggests further reading for more detailed practical guidance on undertaking the different techniques.

The most common causes of primary forest degradation are the over-harvesting of wood and non-timber forest products (NTFPs), overgrazing, and fire. Of these, uncontrolled logging using heavy machinery and poor extraction methods is probably the most important in the humid tropics, adversely affecting soil, remaining trees, water and wildlife. Degraded forests can be classified according to the degree of degradation, as illustrated in Table 9.2.

Table 9.1 *Main restoration objectives and management interventions for different types of degraded forest and other land*

Type of degraded forest/lands	Restoration objectives	Management interventions						Desired outcomes
		PROT	CONS S&W	MGMT REG	ENR PL	PLANT	AGRO-FOR	
Farmland	Restore soil fertility		�©				▦	• Agricultural production systems • Agroforests • Tree plantations • Protected forest • Restored forest cover • Managed secondary forest • Multiple-use forest
	Restore/increase productivity			▦		▦	▦	
	Satisfy subsistence needs						▦	
	Generate income			▦			▦	
	Protect against fire, grazing, wind, etc.	▦						
Riparian areas	Restore/conserve biodiversity	▦			▦	▦		• Protected forest • Restored forest cover • Stable water courses
	Protect streamsides		▦					
	Improve downstream water quality		▦					
Watersheds	Restore/conserve biodiversity	▦		▦	▦			• Protected forest • Restored forest cover • Stable downhill areas
	Prevent and control erosion		▦	▦		▦		

Category	Objective	Intervention columns	Outcomes
Production forests	Stabilize catchments		• Restored/managed forest • Tree plantations • Protected forest
	Restore/increase productivity		
	Restore/conserve biodiversity		
	Protect against fire, illegal logging, poaching, human settlement, etc.		
	Prevent and control erosion		
Protected areas	Generate income		• Protected forest • Restored forest cover
	Restore ecological integrity		
	Restore/conserve biodiversity		
	Increase populations of endangered or threatened species		
	Protect against fire, logging, grazing, etc.		
Mining areas	Restore ecological integrity		• Rehabilitated area • Restored forest cover

Notes: Filled cells indicate management interventions best suited to attain the desired objectives. PROT = protective measures; CONS S&W = conservation of soil and water; MGMT REG = management of natural regeneration (includes practices for retention, induction and tending of natural regeneration); PL = enrichment plantings; PLANT = direct plantations (of mixed or pure species); AGROFOR = agroforests; ENR =

Table 9.2 *Examples of categories of degraded forest in Asia, with restoration options*[2]

Category	Main causes	Main characteristics	Restoration options
Lightly degraded forest	Light logging or light fire	• Still retains the main characteristics of the original forest • Natural regeneration can restore the original forest within a reasonable time	• Natural regeneration: tending of pre-existing wildlings (seedlings, saplings) to improve light conditions (by release cutting or canopy opening)
Moderately degraded forest	Logging, fire or a combination of the two	• Large gaps, generally occupied by pioneer tree species (such as *Macaranga, Homalanthus* or *Glochidion*) • Pioneer trees usually begin to thin out naturally after about ten years of age. Late secondary species or sometimes early primary species develop under them	• Enrichment planting: patch (gap) planting or under-planting in large gaps using mixed species • Natural regeneration if plenty of pre-existing wildlings
Heavily degraded forest	Intensive logging, forest fire or a combination of the two, repeated over time and often with over-extraction of NTFPs	• Most of the primary forest structure has been lost, leaving only a few primary forest species • Large openings occupy at least half the area and are invaded by pioneer weeds, vines, and other secondary forest species • Serious damage to the remaining forest's physical condition (with, for example soil erosion, compaction or impeded water courses) and biological condition (for	• Enrichment planting • Coppice management • Direct tree plantation • Agroforestry systems

Table 9.2 *continued*

Category	Main causes	Main characteristics	Restoration options
		example, lack of seed sources and regeneration of commercial tree species)	
Low-profile hacked forests	Repeated hacking or overlopping (for fuelwood, fodder, poles and small timber), overgrazing and burning	• High susceptibility to fire • Live tree stumps, tree species in shrub form, poles, a few old trees, thickets of shrubs and climbers • Low-height, often dense stands with bad stem form, and affected by pathogens • Hard soil, often sheet- or gully-eroded, usually of low fertility with limited organic matter • Live roots and stumps retain their coppicing ability • Can grow back into high-value secondary forests if over-exploitation is halted	• Site conservation methods • Coppice management • Direct tree plantation • Agroforestry systems

Overview of forest restoration strategies for degraded primary forest

A basic management principle of forest restoration is to use, as much as possible, the natural dynamics already operating in a degraded primary forest stand. Restoration will usually be achieved by the tending of advance growth; it is difficult to induce regeneration from seed since seed sources are often absent and the ground vegetation is usually dense and highly competitive.

Restoration strategies for degraded primary forest will depend on the condition of the forest stand, the objectives of the restoration programme and the resources available. In general, four main (not necessarily mutually exclusive) restoration strategies can be pursued:

1   protection and natural recovery;
2   management of natural regeneration;
3   enrichment planting; and
4   direct plantation.

Each of these strategies entails a series of silvicultural interventions aimed at facilitating the survival and growth of existing regeneration (seedlings, saplings and poles) and also, in the case of the last three, various approaches to planting. This chapter examines these four strategies, and their associated silvicultural interventions, in turn.

Protection and natural recovery

The main restoration objectives of the 'protection and natural recovery' strategy are the conservation of biodiversity and the restoration of ecosystem functioning and, often, commercial productivity. The strategy relies largely on protecting the site from the main disturbance or stress factors and allowing natural colonization and successional processes to occur. This approach is generally most appropriate where the main disturbance or stress factors have been, or can be, controlled effectively, where degradation has not been extensive, and where residual forest patches remain or some advanced forest regrowth is already present. This strategy is sometimes called 'passive restoration' and is particularly suited to situations where the financial resources for FLR activities are limited. However, this strategy does imply a certain level of intervention and investment – in, for example, fire protection measures and the elimination of pests and weeds. The main shortcomings of this approach include the long time required for recovery and the risk that, during this time, other disturbances will emerge to cause further degradation. Nonetheless, this approach is probably the most common and in many situations it is the only one feasible.[3]

Management of natural regeneration

After passive restoration, working with pre-existing natural regeneration is the cheapest and safest means of restoration, provided that sufficient numbers of trees of desirable species are still present; this is usually the case with primary forests that have been only lightly degraded. In more degraded conditions, however, insufficient and unevenly distributed regeneration makes it necessary to resort to more costly silvicultural interventions such as enrichment planting and direct planting. In this section we will refer to silvicultural treatments based on natural regeneration.

In general, silvicultural interventions are necessary in degraded forests designated for timber production to overcome the relative depletion of commercial tree species, to compensate for slow growth rates and to ensure the future commercial timber value of the forest. Options that can be applied, depending on the condition of the forest stand and the forest management objectives, include:

- treatments to improve the growing conditions and yield of advance growth of desirable tree species; and
- treatments to induce and assist the regeneration of desirable tree species.

Desirable tree species may be commercial or potentially commercial timber and/or non-timber species, locally valued species (such as those with household uses or of social, cultural or religious value), and ecologically important species (such as keystone species for wildlife and pollinators).

Treatments to improve the growing conditions and yield of desirable regeneration aim to provide more space for trees of desirable species. Such treatments usually represent the first step towards improving the productivity of the resource and its capacity to meet commercial, social and/or cultural objectives; these are undertaken in two phases. The first phase entails an operation called overstorey removal, in which overmature, defective non-commercial stems (called relicts) are removed, usually by poison-girdling, from the upper levels of the forest canopy. A second phase consists of liberation thinning, a treatment that releases young growth from competition by commercially less-desirable species.

The practical success of silvicultural tending operations depends on:

- the existence of a sufficient number of specimens of the desirable tree species (at least 100 specimens per hectare is generally considered sufficient);
- a more-or-less even distribution of these trees over the entire area; and
- adequate and long-lasting responsiveness of the desirable trees to liberation thinning.

Treatments to induce and assist desirable regeneration are necessary in more heavily disturbed forest, where insufficient or poorly distributed advance regeneration is a major constraint. The first step is to locate and protect any remaining seed trees of the desirable species. These trees are valuable both as seed sources and for the shade they provide. In most cases, the retention of two to six well-formed, reproductively mature trees of the desired species per hectare (or a total of six to ten seed trees per hectare) will probably be sufficient to enhance regeneration. The main criteria to consider in the selection of seed trees include a healthy, well-developed crown and a straight bole free of excessive taper and with no forking below the base of the crown. After selection, the trees should be marked clearly and monitored until seed-fall is completed. All ground vegetation should be cleared within about 20m of the seed tree to facilitate collection.

The most critical need for assisting desirable regeneration is the improvement of light conditions for the seedlings and saplings. Canopy-opening operations, as described above, and treatments in the undergrowth and even at the soil level, as described below, can all be considered.

Cleaning operations (sometimes termed selective/release weeding or underbrushing) aim to reduce competition for resources in order to benefit existing seedlings or seedlings that might become established from seed-fall.

These operations include control measures against aggressive vines and species such as bamboo or undergrowth palms or ferns. Cleaning the undergrowth is a time-demanding and costly intervention and subject to error and carelessness in species identification. A more effective application is to selectively weed below the crowns of a limited number of desirable adult trees prior to seed-fall to specifically enhance the seed germination and seedling establishment of those species.

Thinning is usually applied to juvenile trees of desirable species. It involves the selective removal of saplings or pole-sized stems to favour the growth of the residual stand. This operation is frequently conducted in situations where there is an overabundance of individuals of intermediate size, not all of which can possibly survive until maturity. This sometimes occurs with species regenerating in patches.

Soil-level treatments include controlled burning and mechanical scarification and are particularly useful for species that require mineral-soil seed beds or minimal competition for germination, establishment and subsequent growth.[4]

Enrichment planting

Enrichment planting is defined as the introduction of valuable species in degraded forests without the elimination of the valuable individuals already present. Enrichment may be appropriate in areas where natural regeneration of desired species is insufficient or irregularly distributed, or when the interest is to introduce high-value species that do not regenerate easily.

This silvicultural technique has been used widely in the tropics to supplement a stand's natural regeneration by planting or seeding commercially valuable species, especially where soil characteristics are not conducive to other land uses. It has evolved from simple gap planting to more intensively managed line planting, and even conversion or close planting. The spatial arrangement of the planted seedlings is reflected in the different terms used for enrichment planting:

- *underplanting* – when the artificial regeneration is conducted under a residual stand of non-commercial trees;
- *group planting* – when the seedlings are planted in groups in their final-crop spacing;
- *line planting* – when the trees are planted along cleared lines; and
- *gap planting* – when the seedlings are planted in natural or artificial gaps.

The aim of enrichment planting will depend on the current condition of the forest. Restoring the commercial productivity would be more suitable for slightly to moderately degraded forest. In forests heavily degraded by fire (or in frequently disturbed secondary forests), where only a small number of relatively common species remain, it may be useful to supplement biological diversity in order to hasten the restoration process. For example, it might be necessary to quickly increase the population of several particular plant species that would be unlikely to re-establish well using the passive restoration approach. These might be endangered plant species, plants with large seeds that are poorly dispersed, or plants needed by a particular wildlife species for food or habitat.

Enrichment planting generally consists of transplanting nursery-grown seedlings or wildlings into natural openings in the forest, felling gaps or lines opened specifically for this purpose. The initial plant condition at planting time is a major determinant of success, which emphasizes the importance of obtaining high-quality planting stock from the nursery.

The species planted should be of economic, ecological or social interest. Some important silvicultural characteristics for species suitable for enrichment planting include:

- rapid height growth;
- narrow crown;
- regular flowering and fruiting;
- wide ecological amplitudes;
- tolerance to moisture stress;
- good natural stem form; and
- free of pests and diseases.

The two most common enrichment planting options are line plantings and gap plantings. The choice of method depends primarily on the condition of the forest stand, the restoration objective and the species used. The gap planting method is generally recommended in degraded, over-logged forests, as planting lines are more difficult to open and maintain in these conditions. In dipterocarp forests, line planting is more suitable if the surrounding trees in the stand are small (less than 10cm diameter at breast height).

Line planting consists of opening parallel and equidistant lines in the forest and planting nursery-grown seedlings of commercial species at regular intervals. Planting lines 1.5–2m wide are opened prior to planting by slashing shrubs, ferns and herbs and girdling large unwanted trees. Lines should be opened vertically in a way that allows the seedlings to receive overhead light. The distance between lines is usually 10–20m. Planting in lines is done at a spacing of 2–5m, according to species and size of planting stock. The lines are maintained by cutting trees or branches closing the planting line. All climbers are removed and grasses, herbs and ferns slashed in the planting line.

Gap planting consists of replanting and tending gaps and is the preferred technique in cases where the desired species is relatively light-demanding. A survey should be carried out to determine the location of the gaps and the distribution of already-established seedlings. It is recommended that the planting area diameter be equal to the average height of surrounding overstorey trees. For example, in the case of degraded dipterocarp forests, gaps of about 500m^2 are opened up by cutting all trees. In the gap either a new generation of dipterocarps is planted or the existing ephemeral seedling stock is maintained.

Box 9.1 provides an example of how line planting and gap planting can be combined in a forest restoration programme.

Box 9.1 The INIKEA collaborative project in Sabah, Malaysia

The objective of this project was to improve biodiversity in dipterocarp forests heavily degraded by fire. Under the canopy of a *Macaranga*-dominated forest stand, more than 25 tree species (mainly belonging to the Dipterocarpaceae family, plus some fruit trees) have been planted using two different plantation methods: line planting and gap planting. The latter is cheaper, mainly because the required number of compass lines in line planting is double that for gap planting (where 100 small groups of three seedlings per hectare are irregularly distributed in the forest).

Source: Garcia and Falck (2003)

Many applications of enrichment planting have been unsuccessful mainly because inappropriate species were selected and/or planting and tending practices were inadequate (commonly involving a failure to open the canopy and keep it open and to keep the crowns of the planted trees free of distorting climbers).[5] Clear criteria for successful enrichment planting are provided by Dawkins (described in Weaver, 1996).[6]

Direct plantation

The use of direct tree-planting for restoring degraded primary forests is restricted to localized, more heavily impacted areas (such as areas with harvesting infrastructure like roads and log landings, or open areas invaded by weeds, vines or bamboos). To control erosion and accelerate vegetation recovery in these areas, patches of trees or shrubs can be planted.

In all these cases some previous site preparation is usually necessary, such as weeding and soil ripping to reduce soil compaction. In addition, a good species selection, the use of high-quality planting stock and appropriate planting methods (for instance, mixing with organic debris or fertilizers) are basic requirements to ensure survival and rapid early growth. Direct seeding is also an option.

Box 9.2 illustrates how direct planting can be used to restore log landings and skid trails in logged forests.[7]

Box 9.2 Rehabilitation of log landings and skid trails in southeast Asia

Skid trails and log landings cover a significant proportion of the total area of logged forests. This results in a substantial loss of potentially productive forest. In southeast Asia, for example, it has been estimated that mechanically disturbed sites can cover up to 40 per cent of the logged area. Two techniques can be used for planting dipterocarps in these conditions: direct open planting of seedlings and planting a nurse crop with the subsequent underplanting of dipterocarps.

Open planting is most suitable for skid trails where flanking vegetation provides some remnant canopy and where natural regeneration of pioneer tree species along the skid edges provides organic matter and helps ameliorate the soil. In general, species with drought and heat tolerance and resistance to pests and diseases should be used here. Examples of these species are *Dryobalanops lanceolata, Shorea leprosula* and *Hopea odorata*.

Underplanting is an alternative technique, especially for large open areas. First fast-growing pioneer trees are planted on the site, followed by underplanting with dipterocarp seedlings. Pioneer trees are better adapted to the open conditions of degraded sites and they grow much faster than dipterocarps. Once the dipterocarp seedlings are established, the nurse trees should be thinned to allow increasing amounts of light to reach the seedlings. Examples of fast-growing tree species with potential to act as a nurse crop for dipterocarp seedlings on log landings and skid trails are *Macaranga* spp, *Endospermum malaccense, Octomeles sumatrana* and *Anthocephalus chinensis* among the indigenous species, and the exotic species *Acacia mangium, Albizia falcataria* and *Gmelina arborea*.

Source: Nussbaum and Hoe (1996)

Notes

1   ITTO (2002) defines a degraded primary forest as:

> *a primary forest in which the initial cover has been adversely affected by the unsustainable harvesting of wood and/or non-wood forest products so that its structure, processes, functions and dynamics are altered beyond the short-term resilience of the ecosystem; that is, the capacity of these forests to fully recover from exploitation in the near to medium term has been compromised.*

2   The first three categories are applied to dipterocarp forests and are adapted from Mori (2001). The last category is a special case of heavily degraded forest that is common in large areas of India, Bangladesh, Nepal and Sri Lanka, as described by Banerjee (1995).
3   More detailed guidance on the use of passive restoration can be found in Grieser Johns (1997), Lamb and Gilmour (2003) and Clewell et al (2000).
4   More detailed guidance on the management of natural regeneration can be found in Lamprecht (1989), Wadsworth (1997) and Dupuy (1998).
5   Palmer and Palmer (1989); Weaver (1996).
6   Technical guidance on the use of enrichment planting can be found in Weaver (1987), Palmer and Palmer (1989), Appanah and Weinland (1993) and Dupuy (1998).
7   More technical guidance on direct plantation can be found in Lamprecht (1989), Evans (1992), and Appanah and Weinland (1993).

References and further reading

Appanah, S. and Weinland, G. (1993) *Planting Quality Timber Trees in Peninsular Malaysia. A Review*, Malayan Forest Records No 38, Forest Research Institute Malaysia (FRIM) – German Agency for Technical Cooperation (GTZ), Kuala Lumpur, Malaysia

Banerjee, A. (1995) *Rehabilitation of Degraded Forests in Asia*, World Bank Technical Paper No 270, World Bank, Washington, DC, US

Clewell A., Rieger, J. and Munro, J. (2000) *Guidelines for Developing and Managing Ecological Restoration Projects*, Society for Ecological Restoration, available from www.ser.org

Dupuy, B. (1998) *Bases pour une Sylviculture en Forêt Dense Tropicale Humide Africaine*, CIRAD Document Forafri 4, CIRAD, Montpellier, France

Dykstra, D. and Heinrich, R. (1996) *FAO – Model Code of Forest Harvesting Practice*, FAO, Rome, Italy

Evans, J. (1992) *Plantation Forestry in the Tropics: Tree Planting for Industrial, Social, Environmental, and Agroforestry Purposes*, 2nd edition, Oxford University Press, Oxford, UK

FAO (1998) 'Guidelines for the Management of Tropical Forests – 1. The Production of Wood', FAO Forestry Paper 135, FAO, Rome, Italy

Garcia, C. and Falck, J. (2003) 'How can silviculturists support the natural process of recovery in tropical rain forests degraded by logging and wild fire?', in FAO/ Regional Office for Asia and the Pacific, *Bringing Back the Forests: Policies and Practices for Degraded Lands and Forests*, proceedings of an international conference, 7–10 October 2003, Kuala Lumpur, Malaysia

Grieser Johns, A. (1997) *Timber Production and Biodiversity Conservation in Tropical Rain Forests*, Cambridge Studies in Applied Ecology and Resource Management, Cambridge University Press, Cambridge, UK

ITTO (2002) *ITTO Guidelines for the Restoration, Management and Rehabilitation of Degraded and Secondary Tropical Forests*, ITTO Policy Development Series No 13, ITTO, Yokohama, Japan

Lamb, D. and Gilmour, D. (2003) *Rehabilitation and Restoration of Degraded Forests*, IUCN, Gland, Switzerland and Cambridge, UK and WWF, Gland, Switzerland

Lamprecht, H. (1989) *Silviculture in the Tropics. Tropical Forest Ecosystems and Their Tree Species – Possibilities and Methods for Their Long-Term Utilization,* GTZ, Eschborn, Germany

Mori, T. (2001) 'Rehabilitation of degraded forests in lowland Kutai, East Kalimantan, Indonesia', in S. Kobayashi, J. Turnbull, T. Toma, T. Mori and N. Majid (eds) *Rehabilitation of Degraded Tropical Forest Ecosystems,* workshop proceedings, 2–4 November 1999, Bogor, Indonesia

Nussbaum, R. and Hoe, A. (1996) 'Rehabilitation of degraded sites in logged-over forest using dipterocarps', in A. Schulte and D. Schöne (eds) *Dipterocarp Forest Ecosystems, Towards Sustainable Management,* World Scientific, Singapore

Palmer, J. and Palmer, H. (1989) 'Pre-project study report enrichment planting', report prepared by Japan Overseas Forestry Consultants Association (JOFCA) and Centre Technique Forestier Tropical (CTFT) for ITTO

Peters, C. (1996) *The Ecology and Management of Non-Timber Forest Resources,* World Bank Technical Paper No 322, World Bank, Washington, DC, US

Thomson, L. (2001) 'Management of natural forests for conservation of forest genetic resources', in FAO/Danida Forest Seed Centre (DFSC)/IPGRI, *Forest Genetic Resources Conservation and Management Vol 2: In Managed Natural Forests and Protected Areas (in situ),* International Plant Genetic Resources Institute, Rome, Italy

Wadsworth, F. (1997) *Forest Production for Tropical America,* USDA Agricultural Handbook No 710, USDA, Washington, DC, US

Weaver, P. (1987) 'Enrichment plantings in tropical America', in J. Figueroa, F. Wadsworth and S. Branham (eds) *Management of the Forests of Tropical America: Prospects and Technologies,* proceedings of a conference, San Juan, Puerto Rico, 22–27 September 1986

Weaver, P. (1996) 'Secondary forest management', in J. Parrotta and M. Kanashiro (eds) *Management and Rehabilitation of Degraded Lands and Secondary Forests in Amazonia,* proceedings of an international symposium, Santarem, Para, Brazil, 18–22 April 1993, International Institute of Tropical Forestry, USDA Forest Service, Rio Piedras, Puerto Rico, and UNESCO Man and the Biosphere Program, Paris, France

10

Site-level Strategies for Managing Secondary Forests

Cesar Sabogal

This chapter sets out the possible management objectives and technical options for managing secondary forests as part of an FLR programme. The two main alternative strategies – managing improved fallows without compromising agricultural production and managing forests for production or conservation purposes – are discussed, together with the types of conditions that favour one above the other.

There is considerable ambiguity and confusion in the current use of the term 'secondary forest' both in the literature and in people's perceptions. The term has been applied to numerous types of forests with different characteristics and arising from many different processes. ITTO (2002) defines it as '*woody vegetation regrowing on land that was largely cleared of its original forest cover (i.e. carried less than 10 per cent of the original forest cover)*'.

On the basis of this definition it can be seen that secondary forests:

* result from **significant disturbance to the original primary forest**, with major changes in its structure and composition. Hence, for example, a primary forest that has been selectively logged does not qualify as secondary forest;
* are **distinct from shrubland, grassland or other non-forest vegetation**. Trees are normally defined as woody vegetation more than 3m tall and a forest is defined by FAO as land with more than 10 per cent canopy cover; and
* occupy a successional position **between non-forest vegetation and primary forest**. Over a long period of time, secondary forests can develop similar structures and functions to those of the original forest.

Secondary forests often develop on land abandoned after shifting cultivation, settled agriculture, pasture or failed tree plantations. However, there are some regional differences. In Asia the human-induced disturbances that give rise to secondary forests include severe over-logging (intensive, uncontrolled timber extraction that reduces canopy cover to less than 10 per cent of original cover), shifting agriculture, fire, the rehabilitation of degraded land, and the abandonment of non-forest land uses. In Africa grazing, fire, and fuelwood extraction are the most important disturbance factors that lead to secondary forests.

Secondary forests are often of special economic importance to the rural poor and those who live outside the cash economy because they are usually accessible to local people. They can provide a range of goods to meet immediate livelihood needs, such as timber for housing, fencing and posts, spices, and herbal medicines. Secondary forests are also being increasingly recognized for their value in fallow agriculture, in the industrial timber sector as sources of locally or commercially valuable non-timber forest products (NTFPs), and for the provision of environmental services such as biodiversity conservation, carbon storage, water regulation and erosion control. Boxes 10.1 and 10.2 illustrate some of the diverse values of secondary forests.

Box 10.1 Valuing the biodiversity of secondary forests in the Brazilian Amazon

Secondary forests in Brazil's Bragantina region in the eastern Amazon can be extremely rich in terms of useful species that are an important part of the livelihoods of local people. An ethnobotanical survey carried out in one community helped identify 135 useful plant species providing a wide array of products such as food, tubers, latex, oils, fibres, resins, gums, balsams, condiments, candles and cellulose. The main uses reported were for medicines, food, handicrafts, hunting, construction and other domestic needs. Older second-growth patches could be used and managed for the production of multiple-purpose trees and palms, as well as for medicines, edible fruits, sawnwood and honey. Among the species with the highest potential is *Platonia insignis* (bacuri), whose fruits have a good local and regional demand. This species sprouts easily and abundantly in the area of natural occurrence. There is evidence that farmers are applying simple silvicultural practices, such as liberation thinning to benefit vigorous trees or natural regeneration when the vegetation is still young to promote fruit production. Estimations of fruit production in managed secondary forests show highly competitive financial returns for farmers. Roundwood production from these forests is encouraged by the existence of local markets for uses such as construction and the manufacture of tool handles.

Sources: Smith et al (2001); Rios et al (2001)

Box 10.2 Woodlot management in the Philippines

Private woodlots in the form of managed secondary forests constitute the majority of the forest in the Ifugao province in northern Luzon, Philippines. The woodlots are sited on abandoned shifting cultivation plots and abandoned grasslands. The farmers select the location of the woodlot on the basis of the presence of hardwood seedlings. The boundaries of the woodlot are demarcated by planting fruit trees. Natural regrowth takes place, and as the succession progresses the fast-growing tree species are cut for firewood, releasing the seedlings of desired hardwood species, including some dipterocarps. In addition, some enrichment planting is carried out, mainly of fast-growing reforestation species, fruit trees and rattan. Woodlots are greatly influenced by selective cutting and underplantings. In the two villages studied, approximately 300 plant species were used, cultivated and protected in woodlots by the Ifugao people. The Ifugao use and protect many species for specific purposes. Of the 180 tree species, 77 are considered to produce timber, while 121 are used for firewood. The other species include 36 wild and cultivated fruit tree species and six dipterocarp species. Rattan species occur in the woodlots, and one (*Calamus maniliensis*) is actively cultivated because the Ifugao relish its fruit and attribute medicinal properties to it. A remarkable aspect is the protection of six tree species for their 'water production' properties and the protection of another four tree species because they are believed to be inhabited by spirits.

Source: Klock (1995), adapted by van der Linden and Sips (1998)

The species composition and rate of secondary succession in secondary forests will depend on the amount of site degradation that occurred following clearance. If, for example, the cleared land was subjected to repeated fires or overgrazing, degradation will result in bushy, sparse, low-value vegetation, perhaps with canopy cover as low as 40 per cent. If such extensive degradation has occurred, the degraded secondary forest will require targeted rehabilitation work before any productive management can start.

Typology of secondary forests

Secondary forests can be classified according to their successional stages, vegetation types, ownership patterns, land uses, economic value or other criteria. The following typology, adapted from Chokkalingam and de Jong (2001), is based on the original land use and the nature of the human-induced disturbances that gave rise to the secondary forest. It consists of six categories of secondary forest:

1 **Post-extraction secondary forests**: forests regenerating largely through natural processes after a significant reduction in the original forest canopy (i.e. to less than 10 per cent of the original cover) through tree extraction at a single point in time or over an extended period and displaying major differences in forest structure and/or canopy species composition from that of undisturbed natural forests under similar site conditions in the area.
   forest → harvest → natural regeneration

2 **Swidden fallow secondary forests**: forests regenerating largely through natural processes in woody fallows of swidden agriculture for the purposes of rehabilitating the land and providing products and services for the farmers and/or communities.
   forest → clear & burn → crop → natural regeneration

3 **Secondary forest gardens**: considerably enriched swidden fallows, or less-intensively managed smallholder plantations or home gardens where substantial spontaneous regeneration is tolerated, maintained or even encouraged.
   forest → clear & burn → crop → managed regeneration

4 **Post-fire secondary forests**: forests regenerating largely through natural processes after a significant reduction in the original forest canopy (i.e. to less than 10% of the original cover) caused by fires at a single point in time or over an extended period and displaying major differences in forest structure and/or canopy species composition from that of undisturbed natural forests under similar site conditions.
   forest → fire → natural regeneration

5 **Post-abandonment secondary forests**: forests which regenerate largely through a natural process after the abandonment of alternative land uses such as agriculture or pasture cultivation for cattle.
   forest → alternative land use → abandonment → natural regeneration

6 **Rehabilitated secondary forests**: forests regenerating largely through natural processes on degraded lands. Regeneration might be enhanced by protection from chronic disturbance, site stabilization, water management and planting to facilitate natural regeneration.
   forest → degraded land → rehabilitation + natural regeneration

Overview of management strategies

As secondary forests are usually found on smallholdings or community lands, their management will require an understanding of the role of these forests in farm production systems and within rural communities and the factors that influence land and resource utilization in these areas. These factors, which will also influence any management decisions, include:

*   land tenure (status, access restrictions);
*   farm size and area of agricultural production;
*   site conditions (soil quality and variability, topography, etc.);
*   biological potential (species composition, structure and productivity);

- the market for forest products and services and its accessibility;
- labour availability (family and hired);
- available capital;
- managerial capacity;
- previous knowledge and experience (especially of agroforestry practices); and
- the policy and legal framework (particularly in relation to forest production).

In many situations, secondary forests are fragmented patches in a landscape dominated by non-forest land uses, and management will require a good understanding of the interactions between these uses as well as the associated risks (such as fire or grazing) and opportunities (in terms of forest products and services). Management decisions will therefore need to be taken from a landscape-level perspective and will need to be responsive to changes over time in the biophysical, socio-economic or policy/institutional conditions.

The great variability in the characteristics of secondary forests and the wide geographic distribution of these forests make it difficult to establish general criteria for their management. However, Table 10.1 presents four possible objectives for managing secondary forests and the resulting management systems. The first three systems are production-oriented and range from the improvement of short fallows without compromising agricultural productivity to the use of extended fallow periods and steering management away from the crop–fallow cycle towards longer rotations and forest products. The fourth system aims to maintain secondary forests in the farm/landscape primarily to enhance their protective, environmental or recreational functions and values. This strategy may also be seen as a way to keep a land reserve for future use. The table provides some examples of technological options or management practices for each of these four systems.

Fallow management

The strategy of managing the fallow is particularly attractive in areas with relatively high population density where swidden agricultural systems include short fallow periods, usually no longer than three to four years. Fallow management strategies range from those that aim to sustain annual food cropping to those that focus largely on the production of economically valuable woody vegetation. These strategies correspond to the first two objectives in Table 10.1.

Improved fallow systems aim to accelerate the process of rehabilitation (thereby shortening the length of the fallow period) and satisfy the cash needs and other aspirations of farming households. Agroforestry and silvicultural practices are used to progressively improve or enrich the fallow with desirable trees, shrubs or vines. Box 10.3 describes an improved fallow system in the Peruvian Amazon.

Table 10.1 *Management systems and examples*
of technological options for secondary forests

Management objective	Management system	Examples of technological options/management practices
Increase the efficiency of fallow vegetation to accelerate the recovery of soil productivity for future agricultural use	Short-cycle improved fallow	• leguminous cover crops • organic manure produced outside the field (e.g. animal manure, earthworms) • contour hedgerows and rotational alley-cropping (using short-cycle, semi-perennial species)
Increase the availability of useful products for use in the farming system and to diversify production	Medium-cycle improved/ enriched fallow	• selection and tending of naturally-established, useful (timber and non-timber) tree, palm or shrub species • enrichment with desirable tree species (e.g. those preferred for timber, firewood, fruits, medicine or forage) • multi-strata crops with useful semi-perennial and perennial species
Increase the productivity and value of the secondary forest to generate income through the marketing of timber and non-timber forest products and services	Medium- and long-cycle production forest	• retention and management of seed trees of commercially valuable species • liberation thinning to favour trees of commercial value • canopy opening and undergrowth cleaning to favour establishment of trees of commercial value • soil exposure to favour desirable regeneration • enrichment with commercial tree species (in lines, groups or gaps)
Secure the permanence of the secondary forest to enhance its protective/ environmental/ recreational functions and values	Conservation forest	• protection of useful species for wildlife and as seed trees • collection of wildlings (seedlings, saplings) of desirable species for outplanting in the farm for enriching fallow, high forest, etc. • wildlife management

Box 10.3 Improved fallows in the Peruvian Amazon

The felling of Amazonian floodplain forest during slash and burn agriculture practised by *ribereño* farmers in the Peruvian Amazon is often followed by the abundant natural regeneration of *Guazuma crinita* (bolaina blanca), a medium-density timber tree in high local demand. This natural *taungya* can be managed as an improved fallow, which represents a productive option of growing local importance.

As naturally regenerated secondary forest, the natural *taungya* system helps to stabilize erosion-prone riverbanks and trap fertile sediments, along with providing higher yields for farmers. As the system is based on natural regeneration, establishment costs are lower than for other improved fallows, making it more attractive to farmers. Regeneration should be 're-spaced' after the harvest of agricultural crops (typically crops of maize and rice in successive seasons) with silvicultural thinning in the third year. Maximum mean annual increments of standing volume can reach 20m³/ha/yr. *G. crinita* timber from alluvial forests provides the raw material base for an important micro-industry, supplying large cities with a versatile sawnwood.

Successful management of the improved system by local farmers appears to depend on four main factors:

1   the farmers need to own or possess enough land of adequate quality to allow for fallow rotations of six to seven years;
2   there needs to be adequate seed-rain at the end of the dry season;
3   farmers need to be able to optimize tree–crop density; and
4   farmers need to have the necessary capital, labour and technological knowledge to execute the tending and harvesting operations.

Source: Ugarte (2004)

Farmers use various approaches to create an improved or enriched fallow system. Most commonly they scatter seeds or plant seedlings after harvesting the crops from the site. Box 10.4 lists the characteristics of species most suitable for use in improved fallow systems and provides some examples.

Farmers can also maintain or favour desirable trees and shrubs that have established themselves naturally in the field by taking care not to damage them while cultivating, planting, weeding and harvesting the crop (see Box 10.2). Another option is to seed the species during the agricultural cycle. Here the cultivated crop provides sufficient shade for the seeds to germinate and the emerging seedlings benefit from the relatively high light levels in the agricultural fields, resulting in high growth rates. Once the area is abandoned and secondary succession begins, these seedlings have a size advantage over the recently established and fast-growing pioneers. The larger the size advantage, the more likely it is that these individuals will be the dominant ones in the regrowing vegetation (Peña-Claros, 2001).

Remnant trees in fields may facilitate regeneration in fallows by attracting seed dispersers and creating favourable sites for plant establishment. This will speed up the forest restoration process.

Box 10.4 Species for improved fallows

Characteristics of species used for improved fallows include:

- nitrogen-fixing and/or produce large amounts of organic matter;
- hardy: tolerant of drought and neglect;
- easy to establish;
- removable or short-lived;
- will not resprout continually if cut down – not weedy;
- will not spread to neighbouring crop areas;
- deep-rooted; and
- able to produce useful or marketable by-products such as firewood, poles or edible seeds.

Examples of species used include:

- *Inga edulis* (Inga or icecream bean);
- *Cajanus cajan* (pigeon pea);
- *Crotalaria* spp (sunn hemp);
- *Sesbania sesban* (sesban);
- *Samanea saman* (monkeypod);
- *Gliricidia sepium* (madre de cacao, rata maton);
- *Erythrina* spp (*E. poeppigiana, E. fusca*); and
- *Senna siamea* (pheasantwood).

Source: Wilkinson and Elevitch (no date)

Management for forest products and services

The potential of secondary forests to produce income-earning goods and services will depend on factors such as:

- the biological potential of the resource;
- the existence of markets for the forest products or services; and
- the socio-economic situation and capacity of the landowner/user.

These factors will help determine the potential scope and scale of management strategies – for timber production, multiple-use and conservation.

Managing secondary forests for timber production is likely to be appropriate in only a limited set of situations. Site conditions (including substrate fertility and seed availability) need to favour a high density of fast-growing, light-demanding tree species. And markets need to exist for the kinds of timber products typical of secondary forests. Alternatively, conditions need to exist that enable the regeneration of valuable timber species (from the primary forest) in the secondary stands; these conditions may be present during the early stages of the frontier development process, when contiguous areas of residual forest remain.

Secondary forests may be managed as either *monocyclic* or *polycyclic* systems. Monocyclic (or uniform) systems involve harvesting all the marketable volume of timber in a single felling operation, and the length of the cycle is more or less equal to the rotation age of the species under exploitation. Polycyclic (or selective) systems, on the other hand, involve the repeated harvesting of commercial trees in a continual series of felling cycles. The length of these felling cycles is usually about half the time required for a particular species to reach marketable size. An additional difference between these two systems is that polycyclic systems rely on the existing stock of seedlings, saplings and poles in the forest to produce the harvestable crop for the next felling cycle, whereas monocyclic systems ignore the accumulated growth of these smaller size classes and rely almost entirely on newly recruited seedlings to produce the next crop of trees (Peters, 1996).

When secondary forest stands are dominated by commercially valuable pioneer species, the maintenance of the approximately even-aged condition of the forest through monocyclic silviculture is desirable. This strategy is recommended for pioneer or light-demanding species that require almost complete canopy removal for seed germination or sustainable seedling growth and survival; it is not recommended on degraded or very infertile soils due to its high demands in terms of nutrient exportation. In contrast to polycyclic systems, monocyclic systems create large canopy openings, which may support the invasion of weed species that could impede the growth of commercial tree species and increase management costs. In addition, large canopy removal leaves fragmented secondary forests even more susceptible to fire, so good fire protection techniques need to be applied. Examples of secondary forests managed under monocyclic systems are teak (*Tectona grandis*) forests in Myanmar and okoume (*Aucoumea klaineana*) stands grown mainly in Gabon, Equatorial Guinea and the coastal areas of Cameroon and Congo.

Secondary forests are often suited to multiple-use management, for which a polycyclic system is usually more appropriate. The application of a polycyclic system depends on the presence of desirable, relatively shade-tolerant trees in the lower levels.

Common management interventions based on natural regeneration in polycyclic systems aim to:

- stimulate the advance growth of desirable species;
- induce the establishment of natural regeneration of desirable species; and/
  or
- stimulate the development of resprouts in species with high resprouting
  capacity.

The decision to work only with natural advance growth depends on the quantity and distribution of individuals of the desirable species; a high initial stocking rate is a requirement here. Inducing regeneration from desired species is another option; this depends on the availability of seed sources, which in turn depends on the existence of seed vectors or dispersers (such as animals or wind), the distances involved, the land management practices in surrounding areas, and the phenological characteristics of the species of interest.

The management of resprouts is a common practice in deciduous or dry zones, where more species with high resprouting capacity are found. This method can be applied as simple coppice management or as tree production from seed. In the first case, the forest products (poles and/or firewood) come from thinning and/or the main harvest in relatively short cycles, while in the second case timber production occurs in longer cycles.

When the regeneration of desirable species is concentrated in certain sites and is poor or nonexistent in others, one option is to undertake enrichment planting systematically in lines or groups or more selectively in gaps or patches with favourable site conditions. Planting material can come from the same forest in the form of wildlings or from a nursery. More information on the application of this technique is included in Chapters 9 and 11.

Other silvicultural practices include the removal of litter by soil scarification and controlled burning, which may enhance germination and seedling survival of some timber species in the understorey of secondary forests.

For non-timber forest resources, the focus of treatment is primarily on the seedling and sapling stages. Basic silvicultural operations include selective weeding, liberation thinning and enrichment planting.

References and further reading

Chokkalingam, U. and de Jong, W. (2001) 'Secondary forest: A working definition and typology', *International Forestry Review*, vol 3, no 1, pp19–26

Chokkalingam, U., Smith, J., de Jong, W. and Sabogal, C. (eds) (2001) 'Secondary forests in Asia: Their diversity, importance, and role in future environmental management', *Journal of Tropical Forest Science*, vol 13, no 4, pp563–839

Dubois, J. (1990) 'Secondary forests as a land-use resource in frontier zones of Amazonia', in A. Anderson (ed) *Alternatives to Deforestation: Steps towards Sustainable Use of the Amazon Rain Forest*, Columbia University Press, New York, US

Emrich, A., Pokorny, B. and Sepp, C. (2000) *The Significance of Secondary Forest Management for Development Policy*, TOB Series No FTWF-18e, GTZ, Eschborn, Germany

Finegan, B. (1992) 'The management potential of neotropical secondary lowland rain forest', *Forest Ecology and Management*, vol 47, pp295–321

Fuhr, M. and Delegue, M-A. (1998) 'Silviculture of productive secondary forests in Gabon (Central Africa)', in P. Sist, C. Sabogal and Y. Byron (eds) *Management of Secondary and Logged-Over Forests in Indonesia*, selected proceedings of an international workshop, 17–19 November 1997, CIFOR, Bogor, Indonesia

Guariguata, M. and Finegan, B. (eds) (1998) *Ecology and Management of Tropical Secondary Forest: Science, People, and Policy*, proceedings of a conference held at CATIE, Costa Rica, 10–12 November 1997, Serie Técnica Reuniones Técnicas No 4, CATIE, Turrialba, Costa Rica

ITTO (2002) *ITTO Guidelines for the Restoration, Management and Rehabilitation of Degraded and Secondary Tropical Forests*, ITTO Policy Development Series No 13, ITTO, Yokohama, Japan

Kammesheidt, L. (2002) 'Perspectives on secondary forest management in tropical humid lowland America', *Ambio*, vol 31, no 3, pp243–50

Klock, J. (1995) 'Indigenous woodlot management and ethnobotany in Ifugao, Philippines', *International Tree Crops Journal*, vol 8, nos 2–3, pp95–106

Peña-Claros, M. (2001) *Secondary Forest Succession: Processes Affecting the Regeneration of Bolivian Tree Species*, PROMAB Scientific Series 3, PROMAB, Riberalta, Bolivia

Peters, C. (1996) *The Ecology and Management of Non-Timber Forest Resources*, World Bank Technical Paper Number 322, World Bank, Washington, DC, US

Ríos, M., Martins-da-Silva, R., Sabogal, C., Martins, J., da Silva, R., de Brito, R., de Brito, I., Costa de Brito, M., da Silva, J. and Ribeiro, R. (2001) *Benefícios das Plantas da Capoeira para a Comunidade de Benjamin Constant, Pará, Amazônia Brasileira*, CIFOR, Belém, Brazil

Sips, P. (1993) 'Polycyclic multi-purpose management of tropical secondary rainforests', in J. Parrotta and M. Kanashiro (eds) *Management and Rehabilitation of Degraded Lands and Secondary Forests in Amazonia*, proceedings of an international symposium/workshop, 18–22 April 1993, Santarem, Para, Brazil

Smith, J., Sabogal, C., de Jong, W. and Kaimowitz, D. (1997) *Bosques Secundarios como Recurso para el Desarrollo Rural y la Conservación Ambiental en los Trópicos de América Latina*, CIFOR Occasional Paper No 13, CIFOR, Bogor, Indonesia

Smith, J., Finegan, B., Sabogal, C., Ferreira, M., Siles, G., van de Kop, P. and Díaz, A. (2001) 'Management of secondary forests in colonist swidden agriculture in Peru, Brazil and Nicaragua', in M. Palo, J. Uusivuori and G. Mery (eds) *World Forests, Markets and Policies*, World Forests Volume III, Kluwer Academic Publishers, Dordrecht, the Netherlands

Ugarte, J. (2004) 'Improved natural regeneration-based fallows in lower floodplains of the river Aguaytia, Peruvian Amazon', in *Book of Abstracts*, World Congress of Agroforestry – Working Together for Sustainable Land Use Systems, 27 June–2 July 2004, Orlando, Florida, US

van der Linden, B. and Sips, P. (1998) *Tropical Secondary Forests in Africa and Asia. An Exploratory Compilation of the Literature*, Werkdocument IKC Natuurbeheer nr W-160, IKC Natuurbeheer, Wageningen, the Netherlands

Whitmore, T. (1998) 'A pantropical perspective on the ecology that underpins management of tropical secondary rain forests', in M. Guariguata and B. Finegan (eds) *Ecology and Management of Tropical Secondary Forest: Science, People, and Policy*, proceedings of a conference held at CATIE, Costa Rica, 10–12 November 1997, Serie Técnica Reuniones Técnicas No 4, CATIE, Turrialba, Costa Rica

Wilkinson, K. and Elevitch, C. (no date) 'Improved fallows', *The Overstory*, no 42, available from http://agroforestry.net/ overstory/overstory42.html

11

Site-level Rehabilitation Strategies for Degraded Forest Lands

Cesar Sabogal

Degraded forest land is defined by ITTO (2002) as:

> *former forest land severely damaged by the excessive harvesting of wood and/or non-wood forest products, poor management, repeated fire, grazing or other disturbances or land-uses that damage soil and vegetation to a degree that inhibits or severely delays the re-establishment of forest after abandonment.*

Degraded forest lands are characterized by:

- a lack of forest vegetation (though single or small groups of pioneer trees and/or shrubs may be present);
- low soil fertility;
- poor soil structure (including soil compaction, waterlogging, salinization or other physical and chemical limitations);
- soil erosion;
- recurrent fire and increased susceptibility to fire;
- severe competition, especially from grasses and ferns; and
- a lack of suitable micro-habitants for seed germination or establishment.

The persistent physical, chemical and biological limitations found in degraded forest lands create barriers to natural forest regeneration; an accurate assessment of these factors is key in determining which rehabilitation interventions will be necessary, based on the objectives of the intervention, the landscape context and the available resources.

This chapter focuses primarily on open or denuded forest lands which have been invaded by weeds; it also covers degraded areas affected by soil and water erosion.

The prioritization of degraded forest lands for rehabilitation should take into account the location and condition of these lands, the interests of the stakeholders concerned and the availability of resources for the restoration work. Priority areas comprise degraded areas that are affecting local people's livelihoods, are important for biodiversity and/or are causing environmental problems such as a decline in ecological functions. Thus degraded areas on farmlands, areas of special ecological value or areas that have been subjected to intensive extractive uses (such as mining) would all constitute priority areas for restoration.

This chapter describes four strategies for the rehabilitation of degraded forest lands:

1   protective measures;
2   measures to accelerate natural recovery;
3   measures to assist natural regeneration; and
4   tree-planting.

In addition, the application of agroforestry may be considered a fifth strategy for these areas. This strategy is dealt with in Chapter 12.

Protective measures

A strategy of protective measures usually consists of fire protection and/or erosion control as a means of reverting past degradation and re-establishing the ecological functioning of a forest landscape. In some cases it may be necessary to undertake preliminary repair work prior to the application of silvicultural methods in order to improve soil conditions and/or improve the hydrological functions of the rehabilitation site.

Fire protection

Fire protection is the most important technical challenge in many rehabilitation efforts.[1] A good fire-protection programme begins with an assessment of the climate, areas of high value, areas of high fire risk and priorities for fire protection. It usually includes three components:

1   *fire prevention* to reduce fire risk;
2   *pre-suppression* work to reduce the fuel hazard; and
3   *suppression* of fires once they start.

Fire prevention will require the motivation and involvement of local people. It is important to understand the reasons why people set fires, and this can best be done through dialogue with communities on these reasons and how to address them. For example, communities can be encouraged to establish fuelbreaks around settlements and to develop rules and enforcement systems to prevent farmers' fires from leading to wildfires. In addition, forest restoration practitioners can discuss with community leaders traditional rules and customs about fire and ways in which they can be revived and enforced. The risk of fire accidents and arson can be reduced by awareness-raising efforts and by holding people accountable to the community for fire damage.

Fire pre-suppression aims to reduce fuel hazard by making a site difficult to burn (through fuel-reduction techniques) or by limiting the spread of any fires that do start (through the use of fuelbreaks). Fuel-reduction techniques include:

- **intercropping**: clearing grass between the newly planted trees and replacing it with other crops that do not burn easily; the area must be intercropped and weeded throughout the year to prevent grass from growing;
- **slashing**: cutting and removing grasses or bush vegetation; even if the cut grass is not removed it will still be less flammable than standing grass; and
- **pressing**: pressing the grass low to the ground by trampling or rolling a heavy weight over it; this will make any fires slower-burning with lower flames.

Fuelbreaks are strips of land in which flammable material, particularly grass, has been removed or reduced. Existing man-made firebreaks (such as roads and trails) and natural firebreaks (such as streams, rocky outcrops and gullies) should be used wherever possible (and widened, if necessary), and food and/or wood production should be incorporated into the fuelbreaks. The most obvious place for a fuelbreak is around the borders of high-value areas to protect them from adjacent grasslands. Fuelbreaks should also be placed near the edges of areas (such as grazing grounds) in which fire is used as a management tool or which might otherwise be a source of wildfire. Fuelbreaks may also be located along ridge-tops.

The establishment of live fuelbreaks, or 'greenbreaks', entails the removal of dead plant material and flammable plants (such as *Chromolaena*, grasses and ferns) along the edge of existing forests and shrublands, the control of grazing animals to prevent them causing damage to nearby trees and crops, and the planting of trees at a close spacing (such as 1 x 1m) to achieve rapid crown closure and the early suppression of grass. Species for these greenbreaks need to be easy to establish, able to quickly shade or outcompete invaders (such as *Imperata* grass), and able to survive or resprout if burned. In addition, they should not drop flammable leaves but should retain succulent green foliage throughout the year. Species that have been used as greenbreaks in timber plantations include *Acacia auriculiformis*, *A. mangium*, *Calliandra calothyrsus*, *Gmelina arborea*, *Leucaena leucocephala*, *Syzigyum cumini* and *Vitex pubescens*.

Fire suppression or firefighting is dangerous and difficult, even with good training and equipment. A trained community fire brigade should only attempt fire suppression on small, controllable fires, basing their response on standard fire-fighting procedures. In restoration initiatives the emphasis of fire protection measures should be on fire prevention and pre-suppression.

Erosion control

Erosion control can be achieved economically and effectively through vegetative measures on all but the very steepest slopes (where only solid structures will be able to provide the desired protection and stability). Box 11.1 lists the required characteristics of plants to be used for erosion control. Since a single species rarely possesses all these characteristics it is usually necessary to plant a mixture of trees, shrubs and grasses with complementary characteristics.

Box 11.1 Requirements for plants to be used for erosion control

- ability to grow on degraded and eroded sites;
- rapid development for quick protection;
- deep and widespread root system for good anchorage in the subsoil;
- dense and wide-spreading crown to quickly form a close canopy;
- ease of establishment, preferably by cuttings, stumps or bare-root seedlings;
- high production of litter or nitrogen-fixing, to improve soil conditions;
- ability to withstand physical stresses, such as drought, falling stones and landslides;
- ability to survive when temporarily submerged or in a strong current (important for species to be used in streambank rehabilitation); and
- ability to provide some economic returns from timber, fuelwood, edible fruits or other useful products.

Source: Weidelt (1995)

Planting and sowing is the most commonly used method for slope stabilization. Planting is done in groups and clusters and spacing must be closer than in conventional plantations (on more stable soils and on land with lower gradients). Often a less demanding but site-improving nurse tree is planted first, and then more valuable, but also more demanding species are introduced by under- or inter-planting when the site conditions have improved. In some arid areas, seedlings are planted along contour lines to intercept surface run-off, promote the infiltration of scarce rainwater and control soil erosion. Farmers must be encouraged to lay down contour lines fairly accurately. A good description of practical methods is found in Friday et al (1999).

Another option for controlling soil erosion is to leave unploughed strips of land, 0.5–1.0m wide, along the contour. These natural vegetative strips re-vegetate rapidly with native grasses and weeds, forming stable hedgerows with natural, front-facing terraces. Check-dams, soil traps or diversion canals may also be needed to control water coming from the upslope plot.

Trees can also be planted on terrace edges in order to stabilize the structure and make maximum use of the land. Fruit trees, which need good moisture conditions, can be planted just below the edge of the terrace, where they can benefit from the increased moisture.

Strips of trees along riparian corridors and planted as windbreaks are also important for erosion control. These can also serve as corridors for animal movement and seed dispersal, as well as increasing seedling establishment.

Accelerating natural recovery

Passive rehabilitation through the natural recovery of degraded areas is a viable strategy, but the nature and extent of recovery will be dependent on the ecology and disturbance history of the area and the current condition of the landscape. The character of the biophysical barriers to recolonization will determine the type of rehabilitation measures that can be used. Islands of natural vegetation, however small, are extremely important as sources of seeds, propagules and colonizers. Where such remnants are absent or where a quicker recovery is needed, auxiliary vegetation will have to be introduced by planting or seeding. The four main approaches used in this strategy are outlined below.

Use of remnant trees

Many agricultural areas retain considerable tree cover, whether as individual isolated trees, live fences, windbreaks or clusters of trees. Some of these trees are relicts of the original forest that were left standing when the area was cleared; others have regenerated naturally or been planted by farmers. Isolated or remnant trees are typically retained in pastures and agricultural areas because of their value as sources of timber, fenceposts, firewood and fruits, as shade and forage for cattle, or as sources of organic matter for improving soil fertility, or because their cutting is prohibited by law.

Direct seeding

This approach bypasses the problem of low seed-dispersal rates by introducing the seed directly to the soil by sowing by hand or from tractors or even aircraft. Species most suitable for direct seeding on degraded forest lands have the following characteristics: they produce plenty of seeds, they grow fast in the early stages, they have large seeds with plenty of reserves, they produce a long tap root at an early age and their seeds have a high germinative capacity. Examples of these kinds of species include *Tectona grandis*, *Cassia siamea*,

Leucaena leucocephala, Anacardium occidentale, Albizzia procera and *Dipteryx* spp.

In general, direct seeding gives acceptable results only when the soil is worked (i.e. tilled or scarified) to facilitate root penetration. It is also important that the seeds are covered with a soil layer one to two times the thickness of the seed and, if possible, with a light mulch. To ensure establishment during the rainy season, direct seeding must be carried out at the very beginning of the rains so that the seedlings are already strong enough to survive the heavy rain later in the season. Since success depends on so many variables, small-scale experiments should be conducted first before embarking on a larger operation. Weeding is even more important under this approach than with any other tree-planting method, since the seeds will face competition immediately on germination.

Scattered tree-planting

This approach aims to accelerate succession by increasing the structural complexity that attracts seed- or fruit-dispersing fauna into the degraded landscape from nearby intact forest. One method involves planting small numbers of trees (singly, or in clumps or rows), to form perches for birds. Seedlings are produced from seed-shed below these perch trees and eventually these grow up to form bird perches themselves. The clumps of trees expand and the process continues.

A variant of the scattered tree-planting approach is to use more closely spaced plantings of a small number of species, known as 'framework species', which provide resources such as nectar, fruit or perching sites to attract seed-dispersing birds and bats. Framework species need to be fast-growing, with a dense canopy to shade out weeds, and produce seeds that are easily collected and able to germinate in nurseries. Important groups of framework species include fig trees (*Ficus* spp, Moraceae), legumes (Leguminosae) and oaks and chestnuts (Fagaceae). This method is especially suited to areas close to intact forest that can act as a source of seeds and wildlife. Some maintenance is needed in the early years to ensure that weeds do not dominate the succession.

Patches of dense plantings of many species

The intensive planting of a large number of tree and understorey species is probably the best approach where degraded areas need to be ecologically restored. This would apply to degraded areas surrounding national parks or other protected areas, or those that can be developed as corridors between forest fragments or reserves. The species selected might include fast-growing species able to exclude weeds, poorly dispersed species, species forming mutually dependent relationships with wildlife and, possibly, rare or endangered species that might be present only in small numbers or in small areas. Mostly species from late successional stages should be used, rather than early pioneer species. On the other hand, some short-lived species able to create canopy gaps and regeneration opportunities can be useful.

This approach has the advantage of quickly establishing a large number of species. However, its application can be very expensive due to the need to collect the seeds of and nursery-raise a large number of species and it also requires some knowledge of how to introduce species according to a successional sequence.

Assisting natural regeneration

Assisted natural regeneration (ANR) aims to liberate tree species from competitors, encourage their growth and therefore facilitate their domination over the site. ANR uses the natural regeneration of forest trees (from wildlings and sprouts), assisting it by protecting desired species from fire, controlling weeds and attracting seed-dispersing wildlife. ANR may also include the planting of additional trees (enrichment planting). On appropriate sites, forest cover can be re-established more quickly and cheaply with ANR than with conventional reforestation.

ANR comprises four major component activities:[2]

1 *locating and releasing natural regeneration*: all broadleaf natural regeneration, including specimens hidden amidst the grasses, is clearly located and released by either pressing down (lodging) the grasses, spot brushing or complete brushing;
2 *maintenance operations*: silvicultural treatments such as ring weeding, soil loosening and fertilization should be done as often as necessary until the trees emerge above the grasses;
3 *enrichment planting*: in cases where there are wide gaps between the naturally growing seedlings, the enrichment planting of nursery-grown seedlings (or direct seeding) can be undertaken; and
4 *protection*: the area should be protected from grass fire, by establishing firelines or firebreaks, and from other destructive agents (such as livestock-grazing).

Tree plantations

Tree plantation or reforestation is usually the preferred silvicultural strategy for rehabilitating degraded forest lands. Forest plantations need to be carefully planned, starting with a survey to help identify suitable sites for planting, appropriate soil conservation practices and site preparation methods, sources of seeds and other plant propagules, suitable areas for nurseries, and other important aspects of plantation design, establishment and management.

In many plantation forests, especially those near intact forests and in areas where seed-dispersing wildlife are present, an understorey of native tree and shrub species will develop over time. A large number of species may colonize, leading to a substantial change in the appearance and structure of the plantation. Fostering and managing these understoreys can increase the ecological and social value of the plantations.

The basic principles and technical aspects of plantation establishment and management are well-documented in silvicultural textbooks and manuals (see for example Evans, 1992 and Lamprecht, 1990).

Single-species plantations (monocultures)

Establishing monocultures is usually the main option when the primary objective of rehabilitation is to regain structure, biomass or site productivity for timber production. In this case, trees of a commercially valuable timber species are planted at a high density and exotic species are usually preferred over native or indigenous species, mainly for technical or practical reasons (such as seed availability). However, the particular species selected is usually dictated by ecological considerations (such as site-matching) or social concerns (including the generation of locally valued produce). Monocultures of fast-growing exotic species (such as many pines, acacias and eucalypts) may be useful on severely degraded sites, particularly when these are the only species able to tolerate existing site conditions. On the other hand, timber plantations with native species may serve best for recovering biodiversity and as a long-term investment (for high-quality timber).

Monocultures are limited in their contribution to restoring biodiversity, but they can enable indigenous species to be retained in the region and may benefit wildlife that is adapted to or dependent on them. There are various ways by which the ecological benefits of monocultures can be enhanced, including leaving buffer strips of natural vegetation along streams or rivers or between compartments of the plantation to act as wildlife corridors or protect key watershed areas and establishing a mosaic of monocultures (including some native species) to improve landscape diversity.

Multi-species plantations (polycultures)

In some situations there may be advantages in establishing mixtures of species rather than monocultures, as outlined in Box 11.2. The main problem with mixed plantations is that they are much more complicated to establish and manage.

Mixed-species plantations may take the form of temporary mixtures, where one species is used for a short period as some form of nurse or cover crop, or they may be permanently mixed for the life of the plantation. Care needs to be taken to identify and match complementary species to ensure that the theoretical advantages of polycultures are achieved in practice. The harvesting of fast-growing species has to be considered before the establishment of a plantation and the design of the planting pattern should provide enough space for felling and skidding these trees without damaging others.

Box 11.2 Mixed plantations vs monoculture plantations

Polyculture plantations offer the following advantages over monocultures:

- they offer **enhanced production** because they make better use of a site's above- and below-ground resources;
- they are **less susceptible to pests or diseases** because of microclimate changes that are unfavourable to pests and diseases or because the target trees are dispersed, making it less likely that large pest populations will form;
- they can **combine fast-growing species** (for quick financial returns) **and slower-growing species**;
- they can **act as an insurance policy** when it is difficult to predict the future market value of a particular species;
- they can **better respond to local needs** (e.g. for food, fuelwood and fodder); and
- they can **contribute to landscape biodiversity**.

Source: adapted from Lamb (2003)

Notes

1   The description of fire protection measures here is adapted from Friday et al (1999). General guidelines on fire prevention and control can be found in ITTO (1997).
2   See Dalmacio (1991).

References and further reading

Banerjee, A. (1995) *Rehabilitation of Degraded Forests in Asia,* World Bank Technical Paper No 270, World Bank, Washington, DC, US

Briscoe, C. (1990) *Field Trials Manual for Multipurpose Tree Species,* Multipurpose Tree Species Network Research Series, Manual No 3, Winrock International Institute for Agricultural Development, Arlington, US (also available in Spanish)

Dalmacio, R. (1991) 'Assisted natural regeneration and accelerated pioneer-climax series strategies: Emerging ecological approaches to forestation', in Philippine Council for Agriculture, Forestry and Natural Resources Research and Development – National Program Coordinating Office *Improved Reforestation Technologies in the Philippines,* Book Series No 121/1991

Dubois, J., Viana, V. and Anderson, A. (1996) *Manual Agroflorestal para a Amazônia,* Volume 1, REBRAF – Fundação Ford, Río de Janeiro, Brazil

Evans, J. (1992) *Plantation Forestry in the Tropics. Tree Planting for Industrial, Social, Environmental, and Agroforestry Purposes*, 2nd edition, Oxford University Press, Oxford, UK

Friday, K., Drilling, M. and Garrity, D. (1999) *Imperata Grassland Rehabilitation Using Agroforestry and Assisted Natural Regeneration*, International Centre for Research in Agroforestry, Southeast Asian Regional Research Program, Bogor, Indonesia

ITTO (1993) *ITTO Guidelines for the Establishment and Sustainable Management of Planted Tropical Forests*, ITTO Policy Development No 4, ITTO, Yokohama, Japan

ITTO (1997) *ITTO Guidelines for Fire Management in Tropical Forests*, ITTO Policy Development Series No 6, ITTO, Yokohama, Japan

ITTO (2002) *ITTO Guidelines for the Restoration, Management and Rehabilitation of Degraded and Secondary Tropical Forests*, ITTO Policy Development Series No 13, ITTO, Yokohama, Japan

Kobayashi, S., Turnbull, J., Toma, T., Mori, T. and Majid, N. (eds) (2001) *Rehabilitation of Degraded Forest Ecosystems*, workshop proceedings, 2–4 November 1999, CIFOR, Bogor, Indonesia

Lamb, D. (2000) 'Some ecological principles for re-assembling forest ecosystems at degraded tropical sites', in S. Elliott, J. Kerby, D. Blakesley, K. Hardwick, K. Woods and V. Anusarnsunthorn (eds) *Forest Restoration for Wildlife Conservation*, ITTO – The Forest Restoration Research Unit, Bangkok, Thailand

Lamb, D. (2003) 'Is it possible to reforest degraded tropical lands to achieve economic and also biodiversity benefits?', in FAO, Regional Office for Asia and the Pacific (Bangkok, Thailand) *Bringing Back the Forests. Policies and Practices for Degraded Lands and Forests*, proceedings of an international conference, 7–10 October 2002, Kuala Lumpur, Malaysia

Lamprecht, H. (1989) *Silviculture in the Tropics. Tropical Forest Ecosystems and Their Tree Species – Possibilities and Methods for Their Long-Term Utilization*, GTZ, Eschborn, Germany

Miyawaki, A. (1993) 'Restoration of native forests from Japan to Malaysia', in H. Lieth and M. Lohmann (eds) *Restoration of Tropical Forest Ecosystems*, Kluwer Academic Publishers, The Netherlands

Montagnini, F., González, E. and Porras, C. (1995) 'Mixed and pure forest plantations in the humid neotropics: A comparison of early growth, pest damage and establishment costs', *Commonwealth Forestry Review*, vol 74, no 4, pp306–321

Wadsworth, F. (1997) 'Forest production for Tropical America', in USDA Forest Service (1997) *Agriculture Handbook 710*, USDA, Washington, DC, US

Weidelt, H. (compiler) (1976) *Manual of Reforestation and Erosion Control for the Philippines*, GTZ, Eschborn, Germany

12

Site-level Strategies for Restoring Forest Functions on Agricultural Land

Sandeep Sengupta, Stewart Maginnis and William Jackson

This chapter looks at how agroforestry and other on-farm tree configurations can provide farmers and foresters with a practical means of operationalizing FLR within agricultural landscapes in ways that benefit both human well-being and ecological integrity (thus fulfilling the 'double-filter' criterion of FLR). The chapter starts by exploring the importance of practising FLR in agricultural landscapes and then goes on to provide practical guidance on the main types of agroforestry interventions and key factors for the successful uptake of agroforestry at the landscape level.

Background

The practice of combining on-farm trees with crops and livestock is not a new one; it has been around for several hundreds, even thousands, of years. The more formal science of 'agroforestry', however, emerged only in the 1970s in response to some of the problems of soil fertility, land degradation and deforestation brought about by the advent of modern, intensive and large-scale monoculture farming in the tropics and by the growing interest among researchers and farmers on how tree fallows could be used to improve crop yields.

Agroforestry, as currently defined by the World Agroforestry Centre, is:

A dynamic, ecologically based, natural resources management system that, through the integration of trees on farms and in the agricultural landscape,

> *diversifies and sustains production for increased social, economic and environmental benefits for land users at all levels.* (ICRAF, 2000)

More directly stated, it is a set of land-use practices involving the deliberate combination of trees, agricultural crops and/or animals on the same land-management unit in some form of spatial arrangement or temporal sequence (Lundgren and Raintree, 1982).

Agroforestry has gone through several stages in its development, with its scope widening from providing direct and demonstrable on-site productivity benefits to farmers to providing important off-site public goods or 'forest functions' at the landscape level. These off-site benefits can include improved watersheds, biodiversity conservation and carbon sequestration, as illustrated through the examples of Sukhomajri and Scolel Té in Box 3.2.

Why FLR is important in predominantly agricultural landscapes

It is becoming increasingly clear that forest functions cannot be restored successfully at the landscape level unless sufficient forest restoration efforts are made in the continually expanding agricultural areas that lie outside the current network of forest protected areas and forest production reserves. Agroforestry systems are therefore just as important to FLR as the rehabilitation of degraded forest lands, management of secondary forests, establishment of forest plantations and restoration of primary degraded forests.

It is worthwhile reiterating here that the purpose of FLR is not to return converted forest landscapes (in this case agricultural land) to their original 'pristine' state. Agroforestry and other on-farm configurations, including secondary forests in farm production systems, may not be able (and should not be expected) to act as a substitute for natural forests, but they can offer pragmatic compromises between intensive monoculture farming and natural forest conservation that can yield rich dividends for farmers and foresters alike. While agroforestry may not prevent deforestation in tropical landscapes *per se*, it can nevertheless play a significant role in providing ecological corridors, stepping stones, forest-edge buffers and other habitats for various forest-dependent species, thereby aiding biodiversity conservation within agricultural lands (Schroth et al, 2004). Agroforestry can also be a compromise option for rehabilitating those degraded or deforested forest lands which *de jure* belong to government forest agencies but are under the *de facto* control of poor farmers – and which are often the bone of contention between the two groups (Puri and Nair, 2004).

Agroforestry also offers important livelihood benefits. This is demonstrated by the increasing use of diverse agroforestry practices by rural communities, smallholder investors and individual farmers to gain reliable supplies of wood, non-timber forest products, fuelwood, fodder and shelter and thereby support their agricultural production systems and livelihoods. Such actors also use agroforestry as a coping mechanism to help reduce their production-related risks, particularly in times of drought or crop failure (World Bank, 2005).

Some examples of how agroforestry systems can enhance both human well-being and ecosystem integrity – in other words satisfy the double-filter requirements of FLR – are shown in Box 12.1.

Box 12.1 How agroforestry enhances ecological integrity and human well-being at a landscape scale

- agroforestry systems in Indonesia currently harbour 50 per cent of the plants, 60 per cent of the birds and 100 per cent of the large animals that would normally be found in a natural forest;
- cocoa agroforestry in Cameroon conserves 62 per cent of the carbon found in the natural forest and contains a plant biomass of 304 tonnes/hectare (compared to 85 tonnes/hectare in crop fields);
- in southern Africa, improved fallow agroforestry systems (including species such as *Sesbania sesban*) add soil nutrients equivalent to approximately US$240 worth of chemical fertilizers per hectare. This is particularly important because commercial fertilizers cost two to six times as much in Africa as in Europe and Asia and are rarely affordable to poor farmers;
- in Burkina Faso, the planting of live fences (including *Acacia nilotica, A. senegal* and *Ziziphus mauritiana*) has increased farm incomes by US$40 per year; and
- in Bangladesh it is estimated that 90 per cent of wood used is produced on agricultural land; in India, half of the country's timber is estimated to come from private farm lands.

Sources: World Bank (2002); Sonwa et al (2001); Sanchez et al (1999; as adapted from World Bank, 2005); Garrity (2004)

Main types of agroforestry systems

Agroforestry systems and practices generally fall into two groups – those that are sequential (involving the successive rotation of agriculture and/or livestock production and forestry practices in the form of fallows) and those that are simultaneous (involving combinations of these land uses in some form of spatial arrangement on the same unit of land at the same time) (Leakey, 1996).

Agroforestry systems can also be classified into three broad structural types, namely:

1   *agrisilviculture (tree–crop systems)*: this is a land-use system where both agricultural crops and forest products are produced, simultaneously or sequentially;

2   *silvopasture (tree–grass–livestock systems)*: here the land use is a combination of forestry and livestock management through fodder production and controlled grazing. Silvopasture is a dominant land-use system in the arid zones, which are generally livestock-raising areas. The restoration of *ngitilis* in the Shinyanga region of Tanzania described in Chapter 2 is a good example of community-led silvopasture; and

3   *agrisilvopasture (tree–crop–livestock systems)*: here the land use is a combination of all the above – agriculture, forestry and livestock together on the same land unit, though not always at the same time. Trees provide fodder for animals and nutrients for the crops; crops provide food for the farmers, forage for the animals and organic matter for the soil; and animals provide organic manure that improves soil fertility and enhances crop and tree growth.

In the tropics, agroforestry practices commonly involve improved fallows, shifting cultivation, taungya, home gardens, parkland systems, alley cropping, growing multipurpose trees and shrubs on farmland, boundary planting, farm woodlots, agroforests, plantation/crop combinations, riparian buffers, shelterbelts, windbreaks, conservation hedges, fodder banks, live fences and trees on pasture land (Nair, 1993; Sinclair, 1999, cited in FAO, 2005); some of these are described below.

Agroforestry practices for use in FLR

This section briefly describes some of the practical on-site agroforestry practices and options for implementing FLR in predominantly agricultural landscapes. More detailed guidelines on how to apply these techniques in the field are available in various books and field manuals, including *An Introduction to Agroforestry* by P. K. R. Nair (1993), which is the definitive text on this subject, *Understanding Agroforestry Techniques* by Weber and Stoney (1989), *Manual of Agroforestry and Social Forestry* by Sen et al (2004) and online resources such as the Winrock FACTNet and ICRAF's Agroforestree databases.

Taungya: originally introduced around the mid-1800s, this classical agroforestry system involves the planting of cash or food crops between newly planted forest seedlings on degraded or barren forest lands. This has been preferred particularly when foresters have tried to generate employment and income benefits for poor farmers who do not have any other land for cultivation, in an effort to provide them with direct incentives for engaging in secondary forest restoration activities. Under this system, the landless farmers raise crops while the forest trees are still young. After two to three years, depending on the tree species and spacing, the canopy closes and light-demanding annual crops can no longer be planted. The culminating vegetation is a tree plantation.

The farmers then move to other open forest areas to repeat the process across other sections of the degraded forest landscape. This practice illustrates how agriculture can be practised to benefit the landless poor while contributing to the landscape-level objectives of FLR.

Agroforestry parklands/dispersed trees: the farm/park landscape that covers large parts of the Sahel is a good example of a traditional agroforestry arrangement where trees dispersed in farm fields form an integral part of the cropping system. In semi-arid west Africa, farmers have deliberately managed tree production on their farmland to fulfil their specific needs. Parkland trees provide traditional medicines and basic food commodities – which are of nutritional value for a large number of rural poor – as well as being a major source of wood and non-wood products (Boffa, 2000). Different species are found in these dispersed, park-like stands, depending on local site conditions; among the most familiar are *Acacia albida*, *Butyrospermum parkii*, *Parkia biglobosa*, *Vitellaria paradoxa* and *Borassus aethiopum*. In traditional systems these trees regenerate naturally, so they are distributed fairly evenly across fields in locally random patterns. Where they have been regenerated through human efforts they are usually planted in lines, normally at 10 x 10m spacing (Weber and Stoney, 1989). The important lesson here is that farmers can and do shape their agricultural landscapes to retain and plant trees when there are direct benefits for them, and in doing so the overall ecosystem functionality of the landscape is improved.

Shifting cultivation/improved fallows: shifting cultivation is a traditional agroforestry practice still common in some parts of the world. It involves cyclical agricultural cultivation, whereby farmers cut some or all of the tree crop, burn it and raise agricultural crops for one or more years before moving on to another site and repeating the process. It is ecologically sound provided that the fallow period is long enough to allow the trees to restore soil fertility. To shorten the fallow, trees or shrubs (for example nitrogen-fixing leguminous species) can be proactively planted by local farmers as an FLR intervention instead of allowing the forest to establish itself by natural regeneration. For example, improved fallows such as the cultivation of *Acacia senegal* in Sudan and other semi-arid areas of Africa have been shown to accelerate and enhance the regeneration of soil fertility and produce additional outputs that are of subsistence or commercial value to local farmers (Arnold, 1990).

Boundary planting/borderline trees: borderlines consist of trees, shrubs and grasses that are established to delineate individual farm fields. While providing ecosystem benefits for the wider landscape (for example as components of corridors and stepping stones), borderline trees provide farmers with practical property markers and other useful wood and non-timber forest products. At the same time, since they do not occupy too much space or shade large areas, they do not interfere with regular farming operations (Weber and Stoney, 1989).

Live fences: these typically consist of dense tree or shrub species planted around farm fields to protect them from free-ranging livestock. They are also planted around family compounds and other buildings. This technique differs from borderline plantations in that shrubbier species are used and the shrubs or trees are tightly spaced (0.5–1m) and intensively pruned to maintain a compact, dense barrier. A number of species adapt well for use as live fences. Members of the *Euphorbia* family are particularly good because animals do not browse them. A number of *Acacia* and *Prosopis* species as well as *Leucaena leucocephala*, *Gliricidia sepium* and *Cajanus cajun* are also useful for this purpose (Weber and Stoney, 1989). Establishing live fences can be a constructive FLR intervention in livestock-rearing farming communities, particularly when combined with stall-feeding – free-ranging goats or cattle are thus prevented not only from damaging agricultural crops but also from entering surrounding natural forest lands, which in turn yields additional landscape-level benefits.

Farm woodlots/farm forestry: this usually consists of commercial tree-growing on farm lands, including plantations and secondary forests. Farm woodlots are found in many forms, including timber belts, alleys, block plantations and widespread tree-plantings, and are usually grown by farmers as cash crops to provide an alternative or supplementary source of income. Farm woodlots also provide substantial landscape-level environmental benefits such as wind and salinity control.

Riparian forest buffers: trees, grasses and/or shrubs are planted alongside streams or rivers, often with the aim of providing watershed protection and preventing soil run-off and the pollution of waterways from excess nutrients and chemical pesticides (Beetz, 2002). In this way buffers can help maintain key landscape-level forest functions while providing important on-site benefits to farmers. For example, establishing forest buffers in oil-palm plantations along the Kinabatangan river in Sabah, Malaysia (Box 2.1) helped prevent flooding of the plantations – thereby reducing financial losses for the owners – and also provided habitats/corridors for local endangered wildlife populations. Land around wetlands and alongside drainage canals in irrigation schemes can also provide excellent growing conditions for trees.

Windbreaks or shelterbelts: these are typically linear plantings of trees and/or shrubs (usually several rows wide) established primarily to reduce wind speed, salt or sand intrusion, or snow accumulation and to buffer against extreme temperatures. Integrated with crop or livestock production systems, windbreaks serve to enhance agricultural productivity, improve crop water use, and protect livestock and homesteads; they also provide wider landscape-level ecosystem benefits such as carbon sequestration and habitats for birds and wildlife. In the mountainous Monteverde region of Costa Rica, for example, community planting of 150 hectares of windbreaks has resulted in higher coffee and milk yields by reducing wind damage to pasturelands and livestock and providing wild parakeets with an alternative food source to coffee. These windbreaks, consisting of both indigenous and exotic tree species, are also serving as important biological corridors connecting the remnant forest patches in the area (McNeely and Scherr, 2003).

Home gardens: also known as homestead or mixed gardens, these are usually located close to households and are characterized by a mixture of annual or perennial species including vegetables, fodder, grains, herbs and medicinal plants. They commonly exhibit a multi-strata structure of trees, shrubs and ground flora that recreates some of the characteristics of natural forests (Arnold, 1990) and can be a highly effective FLR measure within agricultural and other managed landscapes in the tropics. Home gardens are widely used across the tropics to supplement outputs from other parts of the farm and can play a key role in diversifying household food sources and in reducing the overall dependence and pressure on forests. They also play a valuable role in conserving biodiversity, including animals: a recent study conducted by IUCN in Sri Lanka found that about 40 per cent of the total inland native vertebrate species were present in traditional home gardens and ricefield managed landscapes (Bambaradeniya, 2003). Tree species commonly planted in home gardens include *Artocarpus hetrophyllus* (jackfruit), *Anacardium occidentale* (cashew), *Cocos nucifera* (coconut), *Azadirachta indica* (neem), *Hevea brasiliensis* (rubber), *Mangifera indica* (mango), *Musa* spp (banana) and *Psidium guajava* (guava).

Complex agroforests: these agroforests usually arise when farmers intercrop a food crop with one or two tall, upper-canopy timber or fruit tree species. Once harvested, the agricultural crops are replaced with other timber and fruit trees with intermediate canopies, and the next time around with low-canopy trees. The end result is one that closely resembles natural forests and often records high levels of biodiversity. Examples of these include the damar (*Shorea robusta*) and durian (*Durio zibethinus*) agroforests of Sumatra and some types of shade-grown coffee and cocoa in west Africa and Latin America (Schroth et al, 2004). Complex agroforests illustrate how simplified agricultural landscapes can be gradually, and incrementally, converted to more biodiversity-friendly forms of agriculture while continuing to provide direct benefits for local farmers.

Alley cropping/hedgerows: this involves planting rows of trees and/or shrubs (either single or multiple species) at a wide spacing to create alleyways where agricultural crops can be planted. The purpose can be to enhance income diversity, reduce wind and water erosion, improve crop production, or improve wildlife habitat or aesthetics. In most alley-cropping systems, trees are planted in straight rows; however, they can also be planted along contours (contour hedgerows) to gain additional soil conservation benefits (Lal, 1995). Leguminous trees, such as *Calliandria calothrysus*, *Leucaena leucocephala*, *Mimosa* spp, *Prosopis cineraria* and *Acacia* spp, are often used in alley-cropping schemes because of their nitrogen-fixing ability, while diverse crops – corn, millet, cowpeas, yams, etc. – are grown in the alleys (Weber and Stoney, 1989). Like all integrated systems, alley cropping requires skilful management and careful planning – particularly in selecting the species combination, since trade-offs can arise between the crops and trees in terms of competition, allelopathic effects, risk of invasives, etc. (Beetz, 2002). Prunings from the trees or shrubs can also be used either as a mulch to increase soil productivity or as fuelwood or fodder.

Key factors for the successful uptake of agroforestry

Despite the many proven benefits of agroforestry systems, their uptake outside those areas where they have been traditionally practised has been generally limited. This, in part, can be explained by the fact that most modern agroforestry techniques have emerged from within research institutions and, being knowledge-intensive, have been successfully adopted mainly in those areas where considerable research support has existed. Further, agroforestry has not yet been fully accommodated within mainstream agricultural incentive or extension schemes, thereby limiting its large-scale uptake by farmers. Over recent years, however, there has been growing recognition of the need to promote agroforestry over a larger geographical area. Successfully implementing FLR in agricultural landscapes will thus depend significantly on making agroforestry an essential component of modern farming.

While there is no single recipe for scaling up, a recent study conducted by Franzel et al (2004) based on case studies of relatively large-scale agroforestry adoption in Asia and Africa identified a number of factors that are needed to successfully scale up agroforestry. Many of these have also been identified in the ITTO guidelines for the restoration, management and rehabilitation of degraded and secondary forests (ITTO, 2002), illustrating the applicability of these guidelines to the restoration and management of forest components within farm production systems as well. They include:

- providing a favourable policy environment and tenure security;
- promoting farmer-centred research and extension;
- offering farmers a range of technical options rather than a specific recommendation;
- ensuring adequate supply and distribution of planting material;
- building local institutional capacity, including through farmer-to-farmer networks;
- improved knowledge sharing and lesson learning; and
- linking farmers to markets and providing value-addition avenues.

The widespread adoption of agroforestry is possible only if farmers benefit from it and support it and when agroforestry practices are developed *with* farmers and not *for* them. As has been observed, unless farmers share substantially in the long-term benefits of forest plantation efforts, the interaction between the 'agro' and 'forestry' components will remain competitive rather than complementary and the aims of FLR in agricultural landscapes will remain unfulfilled (Puri and Nair, 2004). The various agroforestry practices described in this chapter are all closely related and can develop over time from one configuration to another as the trees and other plants mature, and as the needs and intents of the individual farmers change. What is crucial for the successful uptake of agroforestry, and consequently for the success of FLR in agricultural landscapes, is that the process of decision-making, including what trade-offs to make, what planting configuration to adopt and what species to plant, is not planned or imposed from outside in a top-down manner but is instead allowed to be made by the individual farmers or communities in a participatory manner, based on the best information available and through a process of continuous learning and adaptive management.

References and further reading

Sonwa, D., Weise, S., Tchatat, M., Ngongmeneck, B., Adesina, A., Ndoye, O. and Gockowski, J. (2001) *The Role of Cocoa Agroforests in Community and Farm Forestry in Southern Cameroon*, Rural Development Forestry Network, ODI, London, UK

Arnold, J. (1990) 'Tree components in farming systems', *Unasylva*, no 160, FAO, Rome, Italy

Bambaradeniya, C. (2003) 'Traditional home garden and rice agro-ecosystems in Sri Lanka: An integrated managed landscape that sustains a rich biodiversity', in *Proceedings of the International Symposium on Perspectives of the Biodiversity Research in the Western Pacific and Asia in the 21st Century*, 18–19 December 2003, Kyodaikaikan, Kyoto, Japan

Beetz, A. (2002) 'Agroforestry overview: Horticulture systems guide', ATTRA (National Sustainable Agriculture Information Service), Fayetteville, Arkansas, US, available from http://attra.ncat.org/attra-pub/PDF/agrofor

Boffa, J. (2000) 'West African agroforestry parklands: keys to conservation and sustainable management', *Unasylva*, no 200, FAO, Rome, Italy

FAO (2005) *State of the World's Forests 2005*, FAO, Rome, Italy

Franzel, S., Denning, G., Lilleso, J. and Mercado Jr, A. (2004) 'Scaling up the impact of agroforestry: Lessons from three sites in Africa and Asia', in P. Nair, M. Rao and L. Buck (eds) *New Vistas in Agroforestry: A Compendium for the 1st World Congress of Agroforestry, 2004*, Advances in Agroforestry series, vol 1, Springer, New York, US

Garrity, D. (2004) 'Agroforestry and the achievement of the Millennium Development Goals', *Agroforestry Systems*, vol 61, pp5–17

ICRAF (2000) *Paths to Prosperity through Agroforestry. ICRAF's Corporate Strategy 2001–2010*, International Centre for Research in Agroforestry, Nairobi, Kenya

ITTO (2002) *ITTO Guidelines for the Restoration, Management and Rehabilitation of Degraded and Secondary Tropical Forests*, ITTO Policy Development Series No 13, ITTO, Yokohama, Japan

Lal, R. (1995) *Sustainable Management of Soil Resources in the Humid Tropics*, United Nations University Press, Tokyo–New York–Paris

Leakey, R. (1996) 'Definition of agroforestry revisited', *Agroforestry Today*, vol 8, no 1, pp5–7

Lundgren, B. and Raintree, J. (1982) 'Sustained agroforestry', in B. Nestel (ed) *Agricultural Research for Development: Potentials and Challenges in Asia*, pp37–49, International Service for National Agricultural Research, The Hague, The Netherlands

McNeely, J. and Scherr, S. (2003) *Ecoagriculture: Strategies to Feed the World and Save Wild Biodiversity*, Island Press, Washington, DC, US

Nair, P. (1993) *An Introduction to Agroforestry*, Kluwer Academic Publishers, Dordrecht, The Netherlands

Nair, P., Rao, M. and Buck, L. (eds) (2004) *New Vistas in Agroforestry: A Compendium for the 1st World Congress of Agroforestry, 2004*, Advances in Agroforestry series, vol 1, Springer, New York, US

Puri, S. and Nair, P. (2004) 'Agroforestry research for development in India: 25 years of experiences of a national program', in P. Nair, M. Rao and L. Buck (eds) *New Vistas in Agroforestry: A Compendium for the 1st World Congress of Agroforestry, 2004*, Advances in Agroforestry series, vol 1, Springer, New York, US

Sanchez, P., Izac, A-M. and Scott, B. (1999) 'Delivering on the promise of agroforestry', *CGIAR Newsletter*, September, CGIAR, Washington, DC, US

Schroth, G., da Fonseca, G., Harvey, C., Gaston, C., Vasconcelos, H. and Izac, A-M. (2004) *Agroforestry and Biodiversity Conservation in Tropical Landscapes*, Island Press, Washington, DC, US

Sen, N., Dadheech, R., Dashora, L. and Rawat, T. (2004) *Manual of Agroforestry and Social Forestry*, Agrotech Publishing Academy, Udaipur, India

Sinclair, F. (1999) 'A general classification of agroforestry practice', *Agroforestry Systems*, vol 46, pp161–180

Weber, F. and Stoney, C. (1989) *Understanding Agroforestry Techniques*, technical paper no 57, Volunteers in Technical Assistance, Arlington, Virginia, US

World Bank (2002) 'Integrating forests in economic development', in *A Revised Forest Strategy for the World Bank Group*, World Bank, Washington, DC, US

World Bank (2005). 'Agriculture investment note on agroforestry systems', in *Sustainable Natural Resource Management for Agriculture*, Agriculture Investment Sourcebook: Module 5, World Bank, Washington, DC, US, available from www-esd. worldbank.org/ais/index.cfm? Page=mdisp&m=05&p=3

13

Scenario Modelling
to Optimize Outcomes

David Lamb

This chapter looks at some of the key choices to be made when designing an FLR programme and the trade-offs that may be involved in making such choices. It also describes the use of scenario modelling as a tool for making the choices explicit and exploring the different restoration options with stakeholders. Understanding the major trade-offs involved and the kinds of compromises that will need to be sought is crucial to the success of the planned restoration work, considering the different and sometimes conflicting objectives of the many stakeholders involved.

Trade-offs and choices in FLR

Some of the most common issues that will need to be resolved through striking compromises include:

Agricultural area vs forest cover: farmers usually prefer to use the best soils available for food production and to maximize their cash incomes. This can result in forests being cleared even where they are providing critical environmental services to other stakeholders. It may also prevent the restoration of land that is more suited to forestry than agriculture. The trade-off here is between local agricultural production and the regional benefits such as watershed protection and biodiversity conservation that forest cover can best provide. Box 13.1 describes how this trade-off was resolved in a case in Nepal.

Box 13.1 Resolving the agriculture–forest cover trade-off?

One example of how the agriculture–forest cover dilemma has been resolved comes from the Nepal community forestry experience, where farmers had been using degraded common land for open grazing, even though this was the only land available for restoration to provide forests for fuelwood. The compromise that eventually emerged involved farmers altering their livestock management practices from open-range grazing to hand feeding. This drastically reduced the numbers of large animals kept but made common lands available for restoration. One consequence was that fodder needed to be cultivated for the animals. The farmers also insisted that some areas of common land be kept free of trees so they could exercise their oxen to keep them fit for ploughing.

Development or production vs conservation: it is a common view that conservation can be addressed by designating some areas of land as wildlife reserves, thereby leaving the remainder of the landscape to be managed intensively to maximize production. But this view fails to recognize that sustained production in both agricultural and forestry systems depends on the maintenance of key landscape-level ecological processes (such as nutrient cycling or the hydrological cycle) and that these processes, in turn, depend on the maintenance of some degree of biological diversity across the landscape.[1] The very fact that a degraded landscape needs restoration demonstrates how crucial this balance can be. The choices to be made here relate to where and how to reforest in order to re-establish (or maintain) biological diversity and ecological functioning across the landscape.

Species preferences: some stakeholders may prefer to use certain tree species in restoration (for financial or conservation purposes, for example), while site conditions suggest it would be better to initially reforest with a different species, and possibly an exotic one, perhaps because it is more tolerant. In this case the trade-off is between stakeholder preference and the risk of failure (see the example in Box 5.1 in Chapter 5).

Restored forest types: numerous options are available to restore forests in a degraded landscape (as outlined in Chapters 9, 10 and 11) and each will generate a different mix of goods and services. A summary of the main options is provided in Table 13.1. Some options, such as short-rotation pulpwood plantations, provide timber but, possibly, few ecological services. Others, such as enriched secondary forest or multi-species plantations, can provide timber and some ecological services but may be less attractive to investors because of the longer time before commercial benefits are generated. The trade-offs here are between the types of benefits provided, the stakeholders who stand to benefit and the timescale over which the benefits are realized.

Table 13.1 *Simplified summary of goods and services supplied by different restored forest types*

Forest type	Goods	Ecological services	
		Watershed protection	Biodiversity[††]
Natural forest	Moderate[*]	High	High
Secondary forest	Low[**]	Moderate–high	Moderate
Enriched secondary forest	Moderate[**]	Moderate–high	Moderate
Plantation – short rotation	High – but low value	Low	Low
Plantation – long rotation	High – higher value	Low–moderate[†]	Low
Plantation – underplanted with non-timber forest products	High – higher value	Moderate[†]	Low–moderate
Plantation – multi-species	High – higher value	Moderate	Low–moderate

[*]The initial logging yield is often high, though subsequent timber yields may be much lower, especially if the initial logging operation is managed poorly. Yields of other goods can vary considerably over time.
[**]The yield of goods varies with the extent to which the forest has been degraded.
[†]This depends very much on the structure and composition of the understorey layer. Plantations with longer rotations are more likely to acquire such understoreys than those with short rotations.
[††]Biodiversity depends on the type of forest, the size of the forest area, and linkages across the landscape to other forests.

Land tenure: traditional land ownership is often disregarded by central governments. This means that de facto access and use rights may be quite different from formal legal rights. It may be necessary to find some way of resolving, or at least addressing, these differences before anything can happen on the ground. In this case the trade-off is between insisting on strictly legal operations (which may then mean that nothing happens or that restoration projects are sabotaged) and finding some way of accommodating local traditions of access and use rights.

Public interest vs private interest: while farmers manage their land for the benefit of their families, they may pass on some environmental costs (such as increased erosion or the loss of biodiversity) to the broader community. However, constraining farmers' activities on their own lands or insisting that they reforest certain areas would mean that they alone bore the costs of such environmental protection, with the wider community benefiting from this but paying nothing. This mismatch of costs and benefits can be resolved by developing regulations to prevent inappropriate land management, providing some form of compensation to farmers for additional expenses incurred or even making payments for the supply of ecological services such as watershed protection.

Spatial locations: most landscapes contain a mosaic of uses, but there are many ways in which different land uses can be distributed across a landscape. For example, forest might be located in a single large block, with the remainder of the area devoted to agriculture, or it may occur in dozens of small patches. The former situation might be most effective for conserving biodiversity but the latter might be better for regional watershed protection, because strategically located forest patches can protect key erosion-prone sites. Spatial patterns of land use can also affect the extent to which plants and animals can colonize newly restored forest areas. Species-rich understoreys may develop in plantation monocultures if these plantations are near natural forest patches but will not if they are distant. The nature of the landscape mosaic will also determine whether farmers have easy access to non-timber forest products.

Principles for identifying priority sites for restoration

Irrespective of the spatial pattern or degree of degradation, a number of principles can be applied when prioritizing restoration activities to protect resources and enhance productivity across a landscape. The principles listed below assume a good understanding of the given landscape mosaic, including knowledge of the spatial patterns of current land uses and forest types.

- Remaining areas of undisturbed or well-managed natural forest should be protected; such forest should only be cleared after assessment shows this can be justified on economic and/or social grounds and will have minimal environmental impact; plantations established around residual forests are a good way of protecting these from further disturbances.
- Biodiversity within landscapes can be fostered by creating forest linkages or corridors between remaining natural forest areas (see Box 13.2). It is best if these are structurally complex and species-rich, but even monoculture plantations can be useful, especially if natural regeneration produces an understorey beneath the tree canopy.
- Secondary (or regrowth) forest should only be cleared (for agriculture or plantation establishment) after some form of assessment shows that this is justified; in many cases these forests provide important goods and ecological services, especially to local communities.
- Eroding areas (such as hill slopes or river banks) should be stabilized.

- Landscapes should be assumed to be variable – it is rarely the case that a single tree species is the most suitable for plantations at all sites in a landscape (and it is costly to modify sites in order to try to make it so).
- Plantations established to produce sawlogs should use high-value timber species, as these are more likely to retain their value over the length of a rotation. These species usually require longer rotations than firewood or pulpwood species, so the frequency of harvesting and hence the risks of erosion due to site disturbance will be lower.
- Plantations established to produce pulpwood should be located on flatter areas, since the shorter rotations and more frequent harvesting increases erosion risk.

Box 13.2 Building landscape corridors

A common objective of many landscape restoration projects is the establishment of corridors to link separate fragments of intact forest. These corridors enable wildlife and plant species to spread across the landscape and establish themselves more widely, thereby increasing the overall landscape-level diversity and the viability of rare species. Some corridors are small (less than one kilometre long), while others might cover much larger distances. The latter tend to be more significant for biodiversity but are also more difficult to establish. In most cases, the value of corridors depends on the presence of significant remnants of natural forest, so the task is largely one of bridging the gaps between them.

There are two common issues in the establishment of these corridors. The first is whether the links must be made of fully restored forest. In principle, the more structurally complex and species-rich the link is, the more likely it is to be effective. That is, secondary regrowth will usually be better than plantation monocultures. But even monocultures, particularly if grown over long rotations with an understorey, should be able to perform useful roles as biological corridors.

The second issue is whether the corridor must be complete or whether gaps can be permitted. Again, the guiding principle is that a complete link is best but that incomplete corridors may be suitable for many species, provided the gap remaining between forest patches is small. This might be achieved by protecting or planting small forest patches that act as stepping stones between larger areas of forest in the corridor.

Striking compromises

The most crucial trade-offs at the landscape level relate to the *proportion* of the landscape which is devoted to different land uses (agriculture, enriched secondary forests, short-rotation plantations, etc.) and the **spatial location** of each of these different land uses.

Table 13.2 shows four examples of how a landscape might be divided between various land uses. These landscape-level land-use scenarios have been simplified to include just two forms of each major land use. Thus agriculture is regarded as either 'good' (meaning high-quality and productive) or 'poorer' (meaning the land has been abandoned or is used only occasionally for cropping or grazing). Likewise, plantations are simply classified as 'pulpwood plantations' (implying low-quality timbers such as firewood or pulpwood grown on short rotations) or 'sawlog plantations' (implying longer rotations but higher-quality timbers). The existing forests are classified as 'secondary' (in other words recovering after some kind of disturbance) or 'natural' (well-managed and largely intact).

Each of these alternative scenarios has advantages and disadvantages, as outlined below.

Scenario A

Change from current condition: much of the agricultural land (including all the poor-quality farmland) and all the secondary forest are converted to short-rotation plantations.

Advantages: increases forest cover from 15 per cent to 70 per cent and enhances timber production; makes better use of the under-used poor-quality agricultural land and surrounds the small areas of natural forest remnants with plantation forests.

Table 13.2 *Possible coverage (expressed as percentage of land area) of six land uses under four scenarios*

Land-use	Current condition (% cover)	Land-use cover under each scenario			
		Scenario A	Scenario B	Scenario C	Scenario D
Agriculture – good	40	30	40	50	40
Agriculture – poorer	45	0	0	0	0
Total agriculture	85	30	40	50	40
Plantation – pulpwood	0	65	0	0	20
Plantation – sawlog	0	0	0	20	25
All plantation	0	65	0	20	45
Secondary forest	10	0	55	25	10
Natural forest	5	5	5	5	5
Total forest cover	15	70	60	50	60
TOTAL	100	100	100	100	100

Disadvantages: the pulpwood plantations of exotic species contribute little to biodiversity, provide few ecological services, and have been established at the expense of some very good agricultural land and all the secondary forests. Farmers believed they had traditional ownership rights over the farmland, and all the secondary forests were used heavily by local communities for a variety of foods, medicines and ecological services. In the immediate future there will probably be a net decrease in income to local stakeholders because of the decrease in the area of productive agriculture and the loss of goods from the secondary forests.

Scenario B

Change from current condition: forest cover is allowed to regenerate on the low-quality agricultural land, substantially increasing the overall area of secondary forest.

Advantages: increases overall forest cover from 15 per cent to 60 per cent and generates more watershed protection and conservation benefits. The rate at which forest cover increases will depend on the extent of current degradation, the intensity of pressures (such as grazing or fire), and whether there are nearby forest fragments that can act as seed sources. The supply of goods from these secondary forests to local communities might be enhanced by enrichment plantings. Fertilizers and other inputs will boost productivity on the high-quality agricultural lands and so increase the income generated from these areas.

Disadvantages: the recovery rate can be slow if the sites were badly degraded or if intact forest remnants are distant; this can mean there is a risk that the recovery process might be disturbed again by, for example, fires. Only a limited level of biodiversity is likely to develop if the former agricultural sites were heavily degraded or if they covered large, contiguous areas. The initial economic benefits to local communities may be small, especially in the early years when communities lose access to some previously available (though poor-quality) land.

Scenario C

Change from current condition: some poorer agricultural land (10 per cent) is converted into higher-quality agricultural land using inputs such as ploughing and fertilizing. Some of the other poorer-quality agricultural land is used for high-quality sawlog plantations (20 per cent) and the remainder (15 per cent) is allowed to regenerate naturally, adding to the existing secondary forest.

Advantages: increases overall forest cover from 15 per cent to 50 per cent, with some of this being used to provide a buffer around the residual natural forest. There is also an increase in the area of high-quality agricultural land, which will improve community finances. This was flatter land located close to villages.

Disadvantages: this increased forest cover does not address localized erosion problems, which continue to occur on some of the agricultural lands. The plantations all involve long rotations and will take time to provide financial benefits to landholders unless short-term crops such as medicinal plants are grown in the understorey.

Scenario D

Change from current condition: the poorer-quality agricultural land is converted to pulpwood and high-value sawlog plantations. The area of secondary forests remains constant but these are enriched with commercially useful species such as timber trees and medicinal plants.

Advantages: the plantations increase the overall forest cover from 15 per cent to 60 per cent and the value of the secondary forests is enhanced at a faster rate than would have otherwise occurred. The pulpwood plantations have a ready market from a nearby factory and so landowners are guaranteed a regular income. The high-value plantations provide a buffer area around the natural forest remnants.

Disadvantages: poorer farmers dependent on the low-quality agricultural land have lost access to this.

These four examples are highly simplified and do not cover all the trade-off situations described earlier, nor do they show the importance of spatial patterns in determining the outcomes and benefits of the restoration work. However, developing a range of detailed scenarios such as these that are also based on maps showing specific locations for each activity can be a useful tool for exploring options with stakeholders. These scenarios will illustrate which stakeholders are affected by particular options together with the relevant opportunity costs and longer-term benefits. This makes it easier to negotiate trade-offs. One example of such a scenario-setting process is illustrated in Box 13.3.

Developing scenarios and setting priorities

Every situation is different and there is no single 'correct' way to develop FLR scenarios. However, the process will usually involve the following steps:

- understanding the current landscape mosaic and land-use patterns (see Chapter 5);
- arranging meetings with stakeholders and/or their representatives (note that stakeholders will include local land users from within the landscape as well as representatives of the wider community; see Chapter 7);
- defining the existing problems and where they are located (note that different stakeholders may have different perspectives on the identity, nature and location of these problems);
- planning alternative ways of solving these problems, specifying the locations of different restoration options (see Chapter 8);
- developing alternative scenarios to show how compromises might be made to satisfy stakeholders; it is rare to find solutions that satisfy all stakeholders equally;

Box 13.3 Using scenarios to decide land-use options in Papua New Guinea

A pulpwood logging operation planned in Papua New Guinea was going to create a large area of deforested land. A decision needed to be taken on how much of this land should be reforested and how much should be used for different agricultural purposes. This decision was particularly difficult as the region contained eight language groups and 500 land owning clans, most of which practised some form of shifting cultivation. Although the government recognized traditional land ownership patterns, none of the clan land boundaries were formally mapped because access was too difficult. Eight scenarios were developed by a working group made up of clan representatives and landowners as well as government officials. This group undertook extensive field visits. The scenarios that the group developed included spatially specific allocations of land for food production by local communities, riverine protection and nature reserves. Where the scenarios differed was in the amounts and locations of land to be reforested, the types of reforestation employed, and the amounts of land given over to more sedentary forms of agriculture such as large-scale rice farming and cattle-grazing. Discussions eventually led to the creation of a ninth scenario which bridged the difference between two of those already proposed. This option was the one finally adopted by the group. The whole process took several years.

- for each scenario, identifying whether compensation or other incentives are needed to encourage the new land uses; identifying situations where better regulatory controls are needed to prevent practices from causing further degradation;
- consulting stakeholders to assess their preferences for particular scenarios (see Chapter 7);
- establishing what resources (including financial) are available to implement the FLR activities;
- establishing priorities for action: which things must be done first with the resources available, and what can be left for the longer term; and
- planning, implementing and monitoring (see Chapter 14).

Optimizing outcomes – evidence of success

It is difficult to optimize the outcome of an FLR programme, since many ecological processes develop over long periods, during which the economic circumstances (such as commodity prices) and stakeholder objectives can also change. Optimizing FLR outcomes thus becomes another part of the adaptive management approach described in Chapter 4. Some indicators of success might include the following:

- remaining natural forest is protected;
- overall forest cover within the landscape is increased, particularly on steep slopes and riparian strips;
- erosion from steep or other sensitive areas such as river banks is reduced;
- river water quality is improved (especially during periods of heavy rain when the risk of erosion is greatest);
- small forest remnants are enlarged; many become inter-connected by some form of new forest cover (such as secondary forest or plantations);
- native plants and wildlife begin colonizing older plantations;
- land-use boundaries are maintained over time;
- economic productivity (agricultural and forestry) at each site is maintained or improved;
- local communities and other stakeholders recognize the benefits of the initiative and take increasing ownership of it;
- formal codes of practice or informal rules describing good management practice are adopted by stakeholders to prevent future degradation; and
- the need for external financial subsidies or incentives declines.

Note

1   See Chapter 5.

References and further reading

Bennett, A. (1999) *Linkages in the Landscape: The Role of Corridors and Connectivity in Wildlife Conservation*, IUCN, Gland, Switzerland and Cambridge, UK

Hajkowicz, S., Hatton, T., McColl, J., Meyer, W. and Young, M. (2003) *Exploring Future Landscapes: A Conceptual Framework for Planned Change*, Land and Water Australia, Canberra, Australia, available from www.lwa.gov.au

14

Monitoring and Evaluating Site-level Impacts

James Gasana

This chapter outlines the role of monitoring and evaluation (M&E) within an FLR initiative and provides some initial guidance on the process and tools involved in planning and managing M&E work.

Monitoring and evaluating an FLR activity serves to:

- facilitate the efficient and effective use of resources;
- assess progress in the achievement of objectives;
- identify changes in the condition of the forest and the context of the restoration programme;
- support the use of an adaptive management approach; and
- provide information for regular reporting requirements.

In addition to assisting the implementation of the FLR activity, M&E also facilitates learning within the implementation team and the beneficiaries and informs the planning of any follow-up work. While monitoring is the responsibility of the FLR management team, it should be conducted in a participatory manner, involving a wide range of stakeholders.

M&E needs to be built into the FLR initiative from the start. Thus an M&E plan needs to be prepared during the initial planning phase of the restoration work and based on a good understanding of the context of the FLR intervention (see Chapter 4). FLR initiatives establish processes which strengthen local skills and capacities to enable the beneficiaries to continue the activities even after the project ends. These activities face major technical, economic, social, cultural and institutional challenges, requiring an adaptive management approach in order to overcome them, and M&E tools allow practitioners to gather and analyse the lessons learned through 'learning-by-doing'. M&E must therefore be an ongoing process rather than merely an occasional exercise.

In the framework of adaptive management, it provides information to make changes if certain aspects of the project are found to be unrealistic and justify a corrective action. The M&E plan should include:

- a detailed description of the monitoring tasks;
- assignment of the specific responsibilities for these tasks to members of the implementation team;
- a schedule for the M&E activities;
- the set of indicators to be used; and
- identification of the financial and other resources required for the M&E work.

Understanding the context

Logical framework

For the purposes of this chapter, we will assume that the logical framework approach has been used as the basis for planning the FLR activity, in order to show the points at which M&E can be applied. A logical framework sets out the objectives, outputs, outcomes, results, development processes and impacts of an intervention.[1] Each of these terms is described here in turn, and an example of a logical framework based on a hypothetical case is provided in Table 14.1.

The **objectives** of an intervention are defined at two levels: developmental and specific. Development objectives give a general vision of what the intervention is aiming to achieve in the long run, while specific objectives specify the purposes of the intervention, given the available resources and the time-frame set for the restoration work. These two kinds of objectives are defined during the preliminary planning phase, based on an initial assessment of the current conditions and the key issues identified through consultations with the different stakeholder groups.[2]

Outputs are targets to which planned activities contribute directly; they might include specific physical infrastructure, services, studies, identification of boundaries of the target site, community consultations, management plans, maps, training programmes, workshops and publications. The utilization of the outputs delivered by the intervention leads to a set of **outcomes** such as reduced soil erosion, improved access to resources, new or improved markets for resource products, changes in resource-use technology, or the development of new sources of income for communities. The outputs and outcomes constitute the **results** of the intervention. These are the direct changes that are planned in the design of the intervention and are linked to the specific objectives.

Some of the outcomes may be **development processes** produced by the intervention to bring about changes in the landscape and sustain the impacts of the restoration work. These processes, which need to be monitored just as much as the other more concrete outcomes, could include the involvement of key stakeholder groups in the planning and implementation of the FLR work, particular efforts made to involve and benefit women, the poor or other

Table 14.1 *Example of a logical framework based on a hypothetical case*

Project strategy	Measurable indicators	Means of verification	Important assumptions
Development objective: To contribute to sustainable natural resource use by local communities	• New alternative livelihoods for local people	• Socio-economic survey reports • Ex-post evaluation report	• Continued political commitment of the government to the project's objective
Specific objective: To initiate community-based restoration of the degraded forest of the Black Water Forest Reserve (BWFR), Southern Province	• Extent to which illegal activities in BWFR are controlled • Management activities undertaken by the communities	• Project progress reports • Field reports	• There will be a positive response by the community leaders and continued commitment by provincial authorities
Outcome: The management plan for BWFR is under implementation with close collaboration of local communities	• Type of planned activities under implementation • Degree of integration of local communities in management decisions and implementation	• Project progress reports • Evaluation reports	• There will be a positive response by the community leaders
Output 1: Management structure for BWFR established	• By February 2007 the Ministry of Forestry provides project personnel and establishes a project steering committee (PSC)	• Official decisions • Project progress report • Minutes of the first PSC meeting	• Qualified personnel will be available
Output 2: Illegal logging and hunting in BWFR reduced	• By the end of 2009, illegal logging and hunting activities are under control	• Project field reports	• Commitment of public law enforcement agencies
Output 3: Ecological surveys conducted and participatory land-use zoning of BWFR carried out	• Survey reports are published by the end of 2007	• Survey reports and publications • Project progress report	• The management office of the ministry provides adequate backstopping to the survey teams
Output 4: Management plan elaborated and approved	• Management plans approved before the end of 2008 • Act of official approval	• Management plan • Act of official approval	• Continued political commitment of government
Output 5: Activities for community-based resource management initiated	• By the end of 2009, the project is supporting at least one community-based initiative per district	• Project progress reports • Field verification	• There will be a positive response by the community leaders

vulnerable groups, or institution-building processes to strengthen local capacity to capture more sustainable benefits from forest products.

The *impacts* are long-term changes in the condition of the landscape and the biophysical and socio-economic contexts. Impact monitoring consists of the periodic observation of and reflection on the changes caused by an FLR intervention.

Monitoring

Monitoring uses a set of indicators; these are variables that help measure the changes in the socio-economic and environmental conditions of the landscape as a result of the intervention. Indicators are therefore intended to assess the outcomes and impacts of the intervention – not just the implementation effort. An illustrative list of indicators that could be used in a monitoring activity is provided towards the end of the chapter.

The monitoring process requires an initial baseline assessment of the conditions of the landscape and the biophysical and socio-economic contexts against which any future assessments can be compared. The baseline should focus on information that can establish a link between the achievement of the FLR objectives and changes in landscape conditions. The baseline should also attempt to collect information on each of the indicators identified in the M&E plan. If, for some reason, baseline data cannot be collected prior to the implementation of the FLR activity, it should be collected during the earliest possible stage of implementation.

Evaluation

Evaluation is the systematic and objective assessment of ongoing or completed activities, taking into account the design, implementation and impacts of the activities to determine if the objectives have been achieved. The emphasis here is on understanding the reasons for successes and failures and deriving lessons for future phases of the work or to share with others involved in similar interventions elsewhere. Unlike monitoring, evaluation is not a continuous process but is carried out at particular points in time. In general, evaluations are conducted midway through an intervention and at the end of the inter-vention.

A good evaluation of an FLR intervention should aim to assess:

- *the design and strategy*:
  - Does the intervention address the relevant problems/needs of the key stakeholder groups?
  - Were the causes of the problems identified and ranked?
  - Has a clear development objective been defined?
  - Were lessons from similar interventions taken into account?
  - Were the goods and services to be generated by the FLR work adequately described?

- Was a description made of the expected uses of these goods and services?
- Were the benefits deriving from these uses identified?
- Were the major assumptions made regarding implementation success explicitly identified?

- *relevance*:
  - Is the intervention relevant to national goals and strategies (such as environmental protection, biodiversity conservation, sustainable management of natural resources, poverty alleviation, gender equity and clean development)?
- *the achievement of objectives*:
  - Have the planned outputs been achieved? What progress was made towards the intended outcomes and impacts?
  - Have the specific objectives been achieved?
  - Did the intervention contribute to the development objective?
- *implementation of efficiency*:
  - How well has the intervention been managed?
  - Were the most cost-effective options used for the implementation?
  - Are the investment and recurring costs justified?
- *the process*:
  - What consultation, collaboration, joint decision-making or other processes have been undertaken? With which stakeholder groups?
  - What services, and of what quality, have been delivered? To which groups?
  - What changes have resulted from the delivery of these services?
- *sustainability*:
  - Are the outcomes and impacts of the intervention likely to be maintained?
  - Is institutional and local ownership assured?
- *lessons learned*:
  - Are there specific or general lessons to be learned from the experience which are relevant to future stages of the intervention or for similar interventions elsewhere?

Evaluations are usually carried out by independent persons who have not been involved in the design or implementation of the intervention. However, an internal evaluation may also be carried out by the management team and key stakeholders in order to prepare a mid-term or completion report. It is advisable that such an internal effort should be assisted externally to ensure a satisfactory level of objectivity and critical analysis.

Sample indicators for M&E

The following list of indicators is provided as an example of the kinds of variables that can be measured in an FLR activity. However, it needs to be stressed that this is not an exhaustive list, nor will all the indicators given here be relevant to all FLR initiatives. FLR managers will need to draw up a specific list for each FLR programme, based on the prevailing context and conditions.

Process indicators

Stakeholder participation
Key question: Did beneficiaries actively participate in the design, implementation and evaluation of the FLR initiative?
Indicators:
* identification of the right stakeholders and target groups;
* identification of stakeholders' roles in the FLR process;
* inclusion of disadvantaged groups, such as the poor, with attention to gender equity;
* early stakeholder participation in FLR planning;
* participation in implementation and monitoring;
* competence and level of authority of participating stakeholders;
* commitment of participating stakeholders; and
* existence of leadership groups/individuals for community development.

Stakeholder consultations
Key question: What changes in attitude have resulted from consultations on the FLR intervention?
Indicators:
* quality of information shared and how widely it is shared;
* success in implementing the agreed decisions;
* partnerships among stakeholders;
* coordination of stakeholders; and
* institutionalization of consultations to discuss issues and solve problems.

Service delivery
Key questions: Is the initiative reaching the intended beneficiaries, and are they satisfied? What services are provided, to whom, when and for how long?
Indicators:
* stakeholder satisfaction;
* level of access of stakeholders to the advisory and support services;
* level of training of the advisers;
* compliance with the workplans and schedules; and
* extent to which objectives were achieved.

Community needs assessment and dissemination of results
Key questions: How are the needs and perceptions of target groups assessed? Are clear development objectives defined? What are the appropriate interventions?
Indicators:
* information and communication tools produced;

- sensitivity to the needs of weak/disadvantaged groups;
- community satisfaction; and
- level of community ownership of the FLR intervention.

Stakeholder capacity-building
Key question: Is stakeholder capacity being enhanced and, if so, how?
Indicators:
- demonstration actions undertaken;
- implementation of activities associated with project objectives;
- mechanisms for conflict analysis and resolution;
- strength of local self-governing organizations; and
- organizational capacity of women.

Implementation
Key questions: Is the initiative being implemented as planned? Have the target groups and sites been defined? Are beneficiaries involved in evaluation of the activities?
Indicators:
- coordination of key stakeholders;
- incentives for restoration actions; and
- flexibility to adapt as lessons are learned.

Outcome indicators

Strengthened capacity of responsible agencies to support FLR activity
Key question: Was the planning and implementation capacity of the implementing agency or agencies enhanced?
Indicators:
- adequacy of financial resources;
- full-time multidisciplinary staff in charge of pursuing landscape management;
- volume of certified production; and
- level of institutional capacity to sustain the results.

Integrated resource management being undertaken as intended
Key question: Is resource management oriented towards a diversity of goods and services, according to demand and needs?
Indicators:
- approved management plans (forest production, watershed management, protected areas, etc.);
- production diversification (timber and non-timber forest products and environmental services); and
- existence of land-use plans integrating conservation and production.

Diversified landscape components and production
Key question: Is the target area comprising more than one land use and producing a diversity of productions?

Indicators:
- areas of different components of the landscape;
- level of forest productivity;
- levels of resource use;
- diversity of resource users;
- degraded forests are part of land-use plans; and
- non-timber forest production species promoted for sustainable management.

Recovery of ecosystem integrity and restoration of ecological functions
Key question: Were the ecosystem functions restored as intended?
Indicators:
- forest cover;
- species diversity;
- structure of forests;
- areas under natural regeneration;
- planted areas;
- conditions of flora and fauna;
- functions played by restored forests;
- existence of corridors to link forest ecosystems;
- use of local knowledge for landscape management;
- water yield in watersheds;
- improvement of wildlife habitat;
- level of soil erosion;
- frequency of forest fires;
- carbon sequestration; and
- pressure of human activities (domestic animals, crop production, etc.)

Diversified sources of community income
Key question: Do local people have new, sustainable and diversified sources of income?
Indicators:
- availability of resources;
- access to resources;
- provision of wood/fuelwood to communities;
- provision of fodder from plantations;
- value of production;
- number of jobs created;
- jobs which went to targeted groups (women, tribal/ethnic groups, youths, etc.); and
- changes in income.

Economic efficiency and financial viability of FLR area achieved
Key question: Do the economic and financial returns justify the costs?
Indicators:
- costs versus benefits;

- financial resource mobilization from local actors for the sustainability of results; and
- volume and value of locally processed productions.

Participatory M&E of FLR area management taking place as planned
Key question: Is the FLR intervention M&E taking place as planned?
Indicators:
- monitoring tools produced;
- availability of information on ecological and socio-economic dimensions;
- contribution to effective information and reporting; and
- lessons learned.

Managing the M&E process

The monitoring process starts with a description of the information requirements, which will vary according to the specific needs of partners, stakeholders and implementing institutions. The content and format of the reports, as well as the frequency of reporting, will depend on the users of the information and the type of follow-up and decisions to be taken. Managers of the FLR intervention should set up a management information system (MIS) in order to respond to all the information needs in a timely manner. The MIS should streamline M&E as well as special reporting requirements.

The MIS should be set up based on the following questions:

- What are the information needs of the stakeholders and partners?
- What are the priority areas for information?
- What are the sources of the information?
- What are the methods of data collection?
- How are collection responsibilities organized?
- What resources are required and what resources are available?

An expert in a relevant field or an interdisciplinary team of experts should conduct the evaluation; its objective will vary depending on the stage at which it is carried out. The expert or team of experts will need to have clear terms of reference stating the objective of the evaluation, the information base of the assessment, the issues/aspects to be evaluated, the methods to be used, the key stakeholders to meet and involve in the process, and the time-frame. If the existing reports do not all contain all the necessary information for the evaluation, the evaluation may have to start by collecting additional data. This may particularly be the case for some environmental indicators.

Based on the analysis, the evaluation will draw attention to those aspects of the intervention which have not achieved the stated objectives. It will make recommendations on areas that need more attention for the full success of the intervention or for a subsequent phase. The evaluation results should be widely disseminated to contribute information and knowledge and to aid future decision-making.

Notes

1   More information on the use of a logical framework approach can be found in PARTICIP (2000).
2   See Chapter 7 for details of the stakeholder approach.

References and further reading

Guijt, I. and Woodhill, J. (2002) *Managing for Impact in Rural Development: A Guide for Project Monitoring and Evaluation*, International Fund for Agricultural Development (IFAD), Rome, Italy, available from www.ifad.org/evaluation/guide/toc.htm

PARTICIP GmbH (2000) *Introduction to the Logical Framework Approach (LFA) for GEF-financed Projects*, German Foundation for International Development, Bonn, Germany, available from www.undp.org/gef/undp-gef_monitoring_evaluation/sub_undp-gef_monitoring_evaluation_documents (as 'Logframe reader DSE')

15

Getting Started

Stewart Maginnis and William Jackson

The previous 14 chapters have provided the reader with a comprehensive overview of FLR. It should be clear that it is not an alternative approach to forest resource management, but rather has evolved from the theory and practice of sustainable forest management and the ecosystem approach. FLR recognizes that trade-offs are inevitable and stresses the need to balance these in a fair and equitable way. In so doing, FLR shifts the focus from elusive win–wins, where one attempts to maximize the delivery of several important goods and services, to the identification of opportunities for winning more and losing less. A consequence of this pragmatic approach is that management responses need to be locally negotiated and adaptive. Several chapters have been devoted to explaining what this means in practice, from the need to understand what drives change at the landscape level, through advice on how to involve different stakeholder groups and minimize the possibility of conflict, to taking into account how biophysical factors shape the on-the-ground options for restoring forest goods and services at a landscape level. A range of site-level options for delivering FLR has been outlined and some practical approaches have been suggested to help practitioners identify workable trade-offs and monitor whether the resultant configuration of land uses delivers the balance of goods and services that stakeholders desire.

This final chapter draws these points together by providing practical advice on how to start implementing activities within an FLR framework. The key message is that FLR can be initiated through many entry points at varying levels of planning intensity and resource allocation – there is no fixed recipe for success. The restoration of forest goods and services at a landscape level can begin with modest changes in current site-level practices, through a landscape or watershed consultation and planning process, through changes in land-use regulations or policy, or even through a mixture of all three of these approaches.

Imperfect but positive change

Contemplating how to achieve changes at a landscape level can be perplexing. Where and how does one start? Often the list of issues at a landscape level is so long and achieving change is so dependent on additional resources that at first it can appear to be beyond the skills and capacity of any one individual, or even an entire government agency, private-sector company or community. For example, Mansourian et al (2005) argue that any FLR framework ought to have three dimensions – systems analysis, adaptive management and integration – and be able to address 13 separate elements including multi-disciplinary teams, modelling tools and the policy environment. From a theoretical viewpoint it is difficult to challenge the need for such a comprehensive approach, and indeed many of the chapters in this book make similar points. In our view, however, the ability of practitioners to simultaneously address all three dimensions and 13 elements is unlikely to occur in any one place at any one time and thus the theoretically ideal model of FLR needs to be adapted to the practical reality of a given context.

Adopting a pragmatic approach to FLR is also important if practitioners are to avoid the past mistakes of large-scale land-use planning. As Sayer and Campbell (2004) note, the inability to convert integrated land-use theory into results on the ground has led to widespread frustration and disillusionment. The complexities of successful landscape-level planning tend to spark one of two responses: either an endless spiral of consultative meetings from which practitioners progressively disengage, or a reversion to centralized, top-down, prescriptive and expert-driven processes with little if any consideration of the needs of the communities that will be most affected. If FLR is to make a real difference on the ground then it needs to be considered a guiding framework that steers imperfect but positive change rather than as a process seeking a perfect end-state.

This prompts two initial pieces of advice. First, do not worry too much about addressing every aspect of what the various textbooks and articles say FLR ought to include. Concentrate on what is immediately possible rather than putting every condition in place before getting started with practical action. Second, it is perfectly acceptable to change the approach, particularly when faced with new evidence or views that emerge during the FLR process.

The rest of this chapter looks at practical ways in which FLR can be initiated, while recognizing that the general advice offered below does not obviate the need for practitioners to adjust the FLR framework depending on the local biophysical, social, political and institutional contexts.

Finding the starting point

The journey of one thousand miles begins with a single step.

– Chinese proverb

One of the strengths of FLR is that it can be adapted to function on very limited resources. Indeed, many successful examples of landscape-level restoration have begun with site-level action and grown to the broader level. Our point here is that FLR can have multiple entry points – through national policy processes, through formal landscape-level planning approaches, and through individuals taking action at specific locations in the landscape. Whatever the entry point to FLR, it is important to begin by focusing on a rapid **stock-taking** of the situation, particularly the current **context**, the opportunities for **change** and the existing **constraints**, with a view to identifying potential starting points. The process should help reveal the type of practical actions that could be undertaken on individual sites, indicate their contribution within a landscape context and the types of trade-offs that might have to be made, and raise 'red flags' as to the nature of constraints that may be encountered.

Context: establish the existing context by asking three basic sets of questions:

1 What existing actions or initiatives that hold the potential to restore at least some forest-related goods and services are already underway within or close to the area?
2 Will the delivery of those goods and services be limited to the individual site or will they have some off-site impact – either positive or negative?
3 Who is undertaking the actions or initiatives? Who will ultimately benefit? Who is paying the cost (either through direct investment or by foregoing benefits)? Who is interested in forest goods and services?

Addressing these questions will give practitioners a preliminary insight into the potential that exists for promoting FLR. It should also help provide an initial indication of the types of trade-offs that will need to be addressed and who will need to be involved in decision-making processes. Imagine, for example, a situation where 5000 hectares of state-owned secondary forest have been designated for conversion to plantation forests with little local consultation. The possible trade-offs that the practitioner might want to explore could involve retaining natural forest corridors, streamside buffering and areas of set-aside to optimize the ecological integrity of the scheme and negotiating equitable and locally acceptable benefit-sharing arrangements with affected communities to improve social equity.

Change: identifying the opportunities for site-level changes that will have net positive benefits at the landscape level may appear daunting. Yet by applying their existing knowledge and experience in a landscape context, practitioners can begin to separate possible change from impractical idealism. It is particularly helpful to focus on options that improve the delivery of forest functions, rather than trying to speculate on some ideal configuration of forest-related and non-forest land use. To identify the options for change, managers may find it useful to ask themselves three more questions:

1   What forest-related goods and services are missing or inadequate in this landscape? (Remember, however, that this question is only intended to find possible starting points. When you begin the FLR process it is critical that different stakeholder groups speak for themselves.)
2   What actions taken at the local and landscape levels either exacerbate the situation or make a net positive contribution to the delivery of key forest functions?
3   What practical changes can be taken – without compromising the primary management objective of the site – to either reduce the impact of operations or improve the delivery of forest goods and services?

Constraints: even where there is a willingness to change, there may be practical constraints that limit the ability of stakeholders to move forward. These include biophysical limitations such as the loss of top soil, the incremental costs of FLR to landowners or communities, land-use policies that discourage otherwise willing landowners and communities from investing in FLR-oriented activities (for example laws that oblige landowners to maintain forest cover on previously agricultural land once a tree crop has been established), or simply local resistance to adopting new or adapting traditional land-use practice (for example local NGOs may oppose the use of hardy exotic species to rehabilitate degraded sites). It is important to distinguish those constraints that can be overcome via local-level negotiations from those that can only be addressed at a higher policy or institutional level, such as through a change to national legislation. Moreover, a constraint that lends itself to negotiation in one place could be an intractable obstacle in another location.

Two of the key constraints that practitioners will often face are lack of clarity on tenure regimes and use rights and, related to this, ensuring that FLR activities do not benefit solely the better-off while leaving extremely poor people further disadvantaged. Once the FLR process is initiated, it is critical to fully understand how tenure, rights and social position influence landscape dynamics, shape the way that different people regard the importance of forest goods and services, and ultimately determine whether the trade-offs associated with FLR are really fair and equitable.

Initiating an FLR process

The stocktaking exercise may provide some 'best bets' of where to start and can help filter out those options that are likely to run into a dead-end early on. However, FLR has to be responsive to the needs of various stakeholders and cannot be driven solely by the opinion of a small cadre of experts and practitioners consulted during a stocktaking exercise. In practice, the FLR process rarely operates in a stepwise fashion and practitioners need to learn how to alternate adeptly between the following three phases:

1   *investigation* – determining the main stakeholders and interest groups, the geographic scope of the area, the ecosystem services that society desires, and the capacity of the landscape to supply these services;
2   *negotiation and planning* – using participatory approaches and the information obtained during the investigation phase to negotiate plans for achieving human well-being and ecological integrity in the landscape; and
3   *implementation* – stakeholders implement the agreed goals, evaluate results and adapt management over time.

Each cycle should help move the practitioner and the stakeholders progressively towards a better understanding of how site-level actions contribute to landscape-level dynamics and offer up opportunities for complementary FLR interventions on other sites and with other types of land use. For example, as a plantation company begins to modify its practices to protect water courses, it could provide a network of biological corridors and permit access by local communities to collect fuelwood. This interaction with community members leads to a complementary cycle within the FLR process in which options for outgrower schemes are explored and developed. This in turn allows streamside protection to be improved on agricultural smallholdings through the establishment of planted forests. While these smallholding woodlots will be harvested on, say, a ten-year rotation, the net effect is that at any one point in time the majority of drainage systems in the landscape are now better protected from erosion and nitrification.

Investigation

In this phase, the 'best bets' that emerged during the stocktaking exercise are discussed with other key stakeholders; this in turn will almost inevitably lead to changes in the practitioner's initial assumptions and ideas. The first step is to identify the main stakeholders, ascertain their priorities and define the geographic scope of the area within which FLR is to be undertaken.

Chapter 7 of this book provides some excellent guidance on stakeholder approaches. Practitioners should not assume that they already know all that needs to be known about the different stakeholders and in particular should avoid treating poor local communities as a homogenous group. For example, such a community might comprise the 'chronic poor', the 'coping poor' and the 'transient poor', all of whom have very different relationships with, and dependency on, forest resources (Hobley, 2004; Shepherd, personal communication). Realistically, it may take several iterations of the FLR process to fully disaggregate and understand the complex relationships within and between these various social groupings.

Different stakeholders will probably have widely differing views of what constitutes a 'landscape'. As Fisher et al (2005) point out, boundaries are essentially arbitrary – they are defined by people for a particular purpose. In practice, landscapes based on different boundaries defined by different people usually overlap and are often permeable; they can be thought of as a number of landscapes superimposed on each other. So practitioners shouldn't spend too much time trying to arrive at a commonly agreed vision of the actual boundaries of the landscape, especially during the preliminary cycles when the main aim is to identify feasible FLR-type activities for a few individual sites. The real value of teasing out different landscape perspectives is that it allows one group of stakeholders to perceive how another group of stakeholders regards the space within which they live and work; this in turn provides a context for understanding why they prioritize the delivery of certain forest goods and services over others (Fisher et al, 2005). In other words, it provides a framework for helping groups of stakeholders to agree on how to balance the trade-offs inherent in land use. Moreover, encouraging different stakeholder groups to express their own perspectives on the landscape, and how it is changing, can help to broaden the scope of the restoration activities in subsequent cycles of the FLR process.

The investigation phase should also be used to improve the understanding of the structure and function of those biophysical elements in the landscape that are particularly important for the delivery of certain functions. Guidance on this is provided in Chapters 5 and 8.

Another consideration in the investigation phase is the extent to which FLR can improve functional ecological connections between forest fragments, particularly along water courses. Remnant areas of forest that provide real added value in terms of ecosystem functioning should be identified; these may require particular attention during the negotiation phase.

Opportunities for improving human well-being through FLR should be high on the agenda during the investigation phase. While many of these opportunities may only manifest themselves in the long term, modest but quick wins should also be sought. These may include clarifying use and access rights to forest resources, the introduction of innovative benefit-sharing arrangements in return for, say, improved local fire control, or simply ensuring that any income-generating opportunities associated with FLR activities are fully extended to local communities.

Negotiation and planning

Trade-offs are generally an inevitable part of land-use decision-making, especially in modified landscape mosaics of different types of forest and agricultural practice. The challenge is therefore not to avoid trade-offs, but to negotiate them to achieve outcomes that are balanced across the broader landscape. As a wealth of advice and guidance already exists on facilitating and supporting negotiations between different stakeholders, including in Chapter 7, the rest of this section will focus on six key considerations.

I Balancing ecosystem trade-offs or trading-off ecosystem balances?

A natural system will provide a wide range of ecosystem goods and services, while an intensively managed system is capable of supplying large quantities of only one or two specific goods, such as industrial roundwood. A potential trade-off arises: long-term sustainability against short-term profitability. If the choice is made to constantly favour short-term production outputs over ecosystem services, not only will non-market benefits like biodiversity be affected, but the viability of the whole production system itself may ultimately be threatened.

In the 1990s there was much interest in identifying 'win–win' situations in which sustainability could be guaranteed and production maximized. Today, although it is apparent that such opportunities are very rare, many people still feel uncomfortable with the next best alternative of 'winning more and losing less'. Environmentalists object that one cannot trade the survival of a species because extinction, and often local extirpation, are permanent conditions. Human rights advocates argue that it is ethically indefensible to ask the poor to trade their development prospects for global values such as biodiversity that the global community is not prepared to pay for or national values such as economic growth with only the vague promise that someday a small proportion of it will 'trickle down' to their families. The private sector, from large corporations to small family enterprises, argue that in today's globalized world, trading off production against sustainability reduces not only their profitability but also their competitiveness.

It is for these reasons that FLR stresses the importance of ensuring that trade-offs are balanced. In the same way that one does not need to put every single hectare under the same intensive land use to be profitable, one does not need to produce the full range of ecosystem services on every single hectare to be sustainable. It is possible to deliver landscape-level sustainability by balancing different types of land use. Critical to achieving this is the establishment of a negotiating framework to assist stakeholders to ensure that no individual trade-off made at the site level is repeated across entire landscapes to the extent that society experiences a net loss of forest-related goods and services. The different stakeholder groups should be encouraged to be very clear about what goods and services they want to see supplied and what level of supply constitutes the 'bare minimum' and 'good enough'. They should also be encouraged to focus more on the goods and services they want to see delivered rather than whether a particular area of land ought to be set aside for watershed protection or managed for local firewood supply. Such an approach will reduce the risk of polarizing the process.

2 Pluralism is not the same as empowerment

Recently, an increasing emphasis has been placed on pluralist approaches which, put simply, seek to ensure that everyone's opinion is considered equally during negotiations. However, Fisher et al (2005) note that there are serious risks in blindly assuming that pluralist approaches will, by themselves, yield fair outcomes, especially within a landscape context. This is because the stakeholders who are most likely to negotiate effectively are the same ones who already possess power, influence and information and are therefore precisely those who probably already dominate local decision-making. On the other hand, those least likely to be in a position to negotiate for, and achieve,

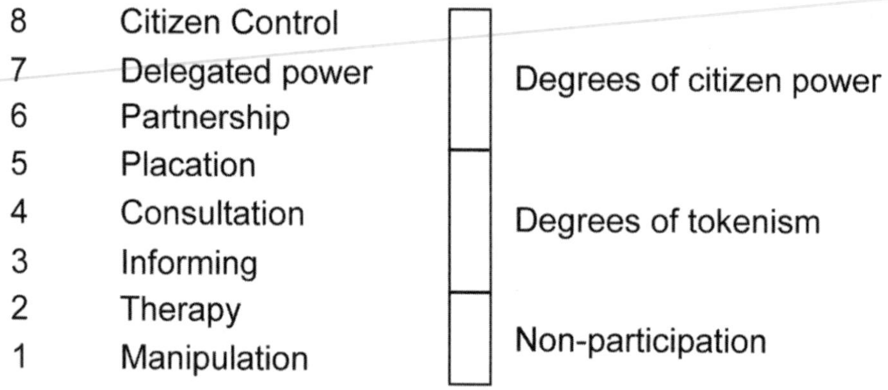

8	Citizen Control	Degrees of citizen power
7	Delegated power	
6	Partnership	
5	Placation	Degrees of tokenism
4	Consultation	
3	Informing	
2	Therapy	Non-participation
1	Manipulation	

Figure 15.1 *'Arnstein's ladder': Eight stages of participation*

Source: Arnstein (1969)

their desired outcomes are the disempowered – in other words the poor and politically marginalized.

'Participation' can play out in many ways, from coercion, through tokenism, to real empowerment, as illustrated by Arnstein's ladder of participation (Figure 15.1). The negotiation and planning phase of the FLR cycle should seek to operate on the higher levels of Arnstein's ladder, empowering stakeholders within the landscape to play an active role in land-use planning and management decision-making, both as individuals and collectively, through partnerships.

3 Facilitate disempowered stakeholders to take a meaningful role
While there may be a genuine desire to avoid tokenistic approaches, it may also be the case that some communities, or particular groups within these communities (such as women), are disinclined to engage in discussions from which they have traditionally been excluded. In such situations it will be necessary to make additional efforts to include such groups, while being sensitive to the reasons they feel unable to engage in the first place. The support of a third party may be required to reach out to these stakeholders, although care must be taken that these intermediaries do not end up becoming the de facto 'voice' of the stakeholders they were intended to support.

The process of empowering is also an iterative one. It takes time to break down barriers of suspicion and prejudice. However, many experiences have shown that empowerment not only pays significant dividends but is often a critical determinant of the success of the FLR intervention. It is important not to mistake disempowerment for an inability to think and act strategically and thus negate the rights of the disempowered to have access to basic information and decision-support tools.

4 Clearly define the role of experts

The rapid development of technology has created many opportunities for improving landscape-level planning and assessment, particularly through the use of computer-based decision-support tools. This presents not only great opportunities but also some major risks. Experts with in-depth technical knowledge but little political awareness of local situations have been known to present one ideal outcome as the 'only' outcome and to preclude any discussion on how to balance trade-offs. However, the technological revolution does allow the redefining of the role of experts, from one of considerable influence in landscape-level decision-making – because they manage the information – to one as facilitator and key provider of knowledge and advice to local decision-makers. Lamb (Chapter 13) and Sayer et al (forthcoming) have been exploring practical approaches that enable poor rural communities to consider different landscape scenarios and on the basis of this define and articulate the sorts of landscape-level outcomes that they hold desirable.

5 Identify and remove constraints to local action

Negotiations should not be about producing a beautiful blueprint for the configuration of an ideal landscape. What practitioners should be looking for is a limited number of practical steps that will help deliver tangible, if modest, change on the ground and help inform the next cycle of investigation, negotiation and implementation. One potentially fruitful area to steer negotiations towards is the identification of practical constraints to implementation that companies, communities or individuals may face.

While the corporate sector and local communities have typically been portrayed as protagonists, recent experience shows that once they have the opportunity to sit together and discuss their individual objectives and perceived threats they can find a significant degree of commonality. Thus encouraging local government officials to remove bureaucratic constraints can prove to be an effective intervention mechanism. For example, Ingles and Hicks (2002) describe an experience in Lao PDR in which the removal of constraints that prevented local communities from participating in the management of local state-owned forests resulted in expanded local markets and private enterprise, significant improvement in the quality of local livelihoods, and a notable restoration of forest quality.

6 Knowing when 'good' is good enough

Shepherd (2004) notes that while long-term goals must be spelled out, unforeseen issues will inevitably lead to the modification of those goals and/ or show new ways to achieve them. This means that the negotiation phase of FLR is not only a means to an end but that the 'end', in other words the implementation phase, will have to be re-examined periodically, reflected upon and adapted and improved as needs be (as described in Chapter 4). Consequently, there will be several occasions during which different stakeholder groups will come together either collectively or bilaterally to learn lessons from what has happened so far and assess how to make further progress. In practical terms, there is little point in attempting to get all the stakeholders involved to agree to a 'master plan' for the landscape. Indeed, we would seriously caution against this approach, particularly since prolonging the negotiation phase until everything is perfect is a recipe for inaction and will likely lead to loss of interest among key participants and their withdrawal from the process.

Implementation

If practitioners are able to help stakeholders agree on some broad goals for the restoration of some of the key functions and identify a small number of supportive actions that could be undertaken at particular sites, then it is time to move into the implementation phase. For example, the very successful establishment of community forestry in Nepal's middle hills, which resulted in the restoration of tens of thousands of hectares of native forest, began with simple actions to plant *Pinus roxburghii*, *Pinus patula* and a number of other exotic species (Jackson et al, 1998).

However, since each situation is different there is little value in trying here to provide a discrete set of steps that should be taken in the implementation phase of an FLR initiative. Rather, we highlight three critical features of the implementation phase that practitioners are advised to keep in mind.

1 ***Learn to adapt***. Adaptive management over time requires the capacity to diagnose the reasons for problems and search for solutions to them (Shepherd, 2004). In our experience, an 'action research' framework (or 'action learning' as Gilmour describes it in Chapter 4) helps to ensure that adaptive management operates as a conscious, guided process of critical reflection and quick response rather than an ad hoc reaction to encouraging success ('do more of the same') or disappointing failure ('try something else'). Action research may sound academic, but it is not. People who share an agreed goal systematically plan, implement and evaluate their actions and from there agree on how to strengthen and improve their future activities. The process has been described for practitioners by Fisher and Jackson (1998).

2 ***You cannot manage what you cannot measure***. Practitioners should make sure that there is some way of assessing whether additional actions are actually making the intended difference. The measurement of landscape-level outcomes is very new, although increasing work is being undertaken to find practical ways forward (e.g. Sayer et al, forthcoming). When establishing benchmarks, it is important to ensure that they can be used to reflect the desired change rather than simply recording whether an agreed activity has taken place.

3 ***Develop bifocal vision***. One of the conundrums of implementing site-level action with landscape-scale ambitions is that one can quickly get caught up in local detail and the need to get results on the ground and forget about moving to the next level. Alternatively, it is possible to worry constantly about the 'bigger picture' and fail to undertake the first practical

steps. It is not a question of what is more important – the longer term and larger scale or the immediate and local-level activities. Both are equally important, and practitioners need to develop the skills to broaden the context of their work while making sure that progress is happening on the ground. A good example of this is how a local movement, Casa Pueblo, which was established by local communities in Puerto Rico to oppose the opening of a large-scale mine, managed to shape forest policy and make a significant contribution to that country's remarkable restoration of its forest resources over twenty years (Massol González et al, 2006).

Conclusion

Many regions are disadvantaged because they have lost key forest-related goods and services. Whether these key functions are to be regained will depend not only on the protection and management of the remnant resource, but also on finding practical ways to put forests and trees back into landscapes. Many practising foresters will also realize the limitations of site-level interventions and understand the rationale for moving towards a more comprehensive landscape-level approach that reflects and helps balance the values and opinions of various stakeholder groups. However, for many practitioners the major challenge will still remain how to take those first critical steps.

Thanks to the increasing interest in FLR, there is a growing body of case studies available for insights into the process and several publications that provide guidance. However, FLR is only a framework, not a strictly defined approach, and practitioners should feel empowered to borrow from past experience and lessons in order to create their own restoration methodologies. These, in turn, may provide inspiration to others. We hope that this book proves helpful in encouraging practitioners to see the potential of FLR in their own situations and provides them with the confidence to take the first important steps towards restoring modified and degraded landscapes for the benefit of both people and nature.

References and further reading

Arnstein, S. R. (1969) 'A ladder of citizen participation', *Journal of the American Institute of Planners*, vol 35, pp216–224

Bennett, G. (2004) *Linkages in Practice: a Review of their Conservation Value*, IUCN, Gland, Switzerland and Cambridge, UK

Fisher, R. and Jackson W. (1998) *Action Research for Collaborative Management of Protected Areas*, Workshop on Collaborative Management of Protected Areas in the Asian Region, 25–28 May 1998, Sauraha, Nepal

Fisher, R. J., Maginnis, S., Jackson, W. J., Jeanrenaud, S. and Barrow, E. (2005) *Poverty and Conservation: Landscapes, People and Power*, IUCN, Gland, Switzerland and Cambridge, UK

Hobley, M. (2004) *The Voice-Responsiveness Framework: Creating Political Space for the Extreme Poor*, Chars Organisational Learning Paper 3, DFID, London, UK, see www.livelihoods.org

Ingles, A. and Hicks, E. (2002) 'A review of the context for poverty reduction and forest conservation in Lao PDR and a preliminary look at IUCN's NTFP Project', presentation given at 3I-C Project Core Team Meeting, 24–26 June 2002, Khao Yai, Thailand, unpublished

Jackson, W. J., Tamrakar, R. M., Hunt, S. and Shepherd, K. R. (1998) Land use changes in two middle hill districts of Nepal, *Mountain Research and Development*, vol 18, no 3

Massol González, A., González, E., Massol Deyá, A., Deyá Díaz, T. and Geoghegan, T. (2006) *Bosque del Pueblo, Puerto Rico: How a Fight to Stop a Mine Ended up Changing Forest Policy from the Bottom-up*, Policy that Works for Forests and People no 12, IIED, London, UK

Maginnis, S., Jackson, W. J. and Dudley, N. (2004) 'Conservation landscapes: Whose landscapes? Whose trade-offs?' in T. McShane and M. Wells (eds) *Getting Biodiversity Projects to Work: Towards More Effective Conservation and Development*, Columbia University Press, New York, US

Mansourian, S., Aldrich, M. and Dudley, N. (2005) 'A way forward: Working together toward a vision for restored forest landscapes', in S. Mansourian, D. Vallauri and N. Dudley (eds) *Forest Restoration in Landscapes: Beyond Planting Trees*, Springer, New York, US

Millennium Ecosystem Assessment (2005) *Ecosystems and Human Well-being: Synthesis*, Island Press, Washington, DC, US

Sayer, J. and Campbell, B. (2004) *The Science of Sustainable Development: Local Livelihoods and the Global Environment*, Cambridge University Press, Cambridge, UK

Sayer, J., Campbell, B., Petheram, L., Aldrich, M., Ruiz Perez, M., Endamana, D., Nzooh, Z., Defo, L., Mariki, S., Doggart, N. and Burgess, N. (forthcoming) *Assessing Environment and Development Outcomes in Conservation Landscapes*, manuscript in preparation

Shepherd, G. (2004) *The Ecosystem Approach, Five Steps to Implementation*, IUCN, Gland, Switzerland and Cambridge, UK.

About the authors

James K. Gasana is a natural resource management specialist working with the Swiss Foundation for Development and International Cooperation, Intercooperation. He has extensive field experience in natural resource management and integrated rural development projects, in national planning of the rural sector, and in managing negotiation processes to settle socio-political conflicts.

Don Gilmour is a forester who worked for eight years in Nepal developing community forestry modalities, followed by five years as head of IUCN's Forest Conservation Programme in Switzerland. He is currently based in Australia and working as a private consultant on a range of forestry and environmental management issues.

William Jackson is the Director of IUCN's Global Programme. He has extensive field experience in ecosystem conservation and management at the global level and of the situations in Asia, Australia and Africa in particular. He has worked with many governments and IUCN partner organizations in devising forest conservation programmes and policies and in evaluating conservation and rural development projects.

Wil de Jong is Professor of the Japan Center for Integrated Area Studies at the National Museum of Ethnology in Osaka, Japan. His work focuses on forest governance and smallholder natural resource management. He has conducted field research in Bolivia, Peru, Zimbabwe, Indonesia and Vietnam.

Trikurnianti Kusumanto is a social science researcher at CIFOR in Bogor, Indonesia, with research interests that include stakeholder collaboration and social learning related to forests. At present she is also conducting research for her doctorate concerning public engagement with forest science and policy.

David Lamb is a tropical forest ecologist based at the University of Queensland, Australia. He is also a member of the IUCN Commission on Ecosystem Management. He is a co-author with Don Gilmour of *Rehabilitation and Restoration of Degraded Forests*, published by IUCN in 2003.

Stewart Maginnis is head of IUCN's Forest Conservation Programme. He has 14 years of practical forest restoration and management experience in Tanzania, Sudan, Ghana, Costa Rica, Mexico and the UK. His areas of interest include FLR, community forest management, the role of forest management in biodiversity conservation and forest policy.

Jennifer Rietbergen-McCracken is a freelance researcher/writer specializing in conservation and sustainable development issues. She is based in Geneva, Switzerland.

Cesar Sabogal is a forest scientist working for CIFOR at its regional office in Belém, Brazil. His research interests include natural forest management and silviculture and forest inventory and monitoring. He was a contributor to the ITTO *Guidelines for the Restoration, Management and Rehabilitation of Degraded and Secondary Tropical Forests* published in 2002.

Alastair Sarre is a freelance writer and editor on forestry and the environment. He is based in Adelaide, Australia.

Sandeep Sengupta currently works as a project officer in the IUCN Forest Conservation Programme in Gland, Switzerland. He has worked on a variety of forest restoration, rural livelihood and natural resource management issues in both government and non-government agencies in India and other parts of South Asia and holds master's degrees from the London School of Economics and the Indian Institute of Forest Management.

Acronyms and Abbreviations

ANR	assisted natural regeneration
CGIAR	Consultative Group on International Agricultural Research
CIFOR	Center for International Forestry Research
CTFT	Centre Technique Forestier Tropical
DFSC	Danida Forest Seed Centre
FAO	Food and Agriculture Organization of the United Nations
FLR	forest landscape restoration
FRIM	Forest Research Institute Malaysia
GDP	gross domestic product
GIS	geographic information systems
GPFLR	Global Partnership on Forest Landscape Restoration
GTZ	German Agency for Technical Cooperation
HASHI	Hifadhi Ardhi Shinyanga
ICDP	integrated conservation and development project
ICRAF	International Centre for Research in Agroforestry
IFAD	International Fund for Agricultural Development
IIED	International Institute for Environment and Development
IPGRI	International Plant Genetic Resources Institute
ITTO	International Tropical Timber Organization
IUCN	International Union for the Conservation of Nature and Natural Resources (World Conservation Union)
JOFCA	Japan Overseas Forestry Consultants Association
M&E	monitoring and evaluation
MIS	management information system
NGO	non-governmental organization
NTFP	non-timber forest product
UNESCO	United Nations Educational, Scientific and Cultural Organisation
USDA	United States Department of Agriculture
WWF	World Wide Fund for Nature

Glossary

Action learning – a process of learning whereby a group of people with a shared issue or concern collaboratively, systematically and deliberately plan, implement and evaluate actions.

Adaptive management – an approach to the management of complex systems based on incremental, experiential learning and decision-making, supported by active ongoing monitoring of and feedback from the effects of outcomes of decisions.

Advance growth – tree seedlings and saplings that have become established naturally in a forest and that will form the basis of natural regeneration.

Agroforestry – a dynamic, ecologically based, natural resources management system which, through the integration of trees on farms and in the agricultural landscape, diversifies and sustains production for increased social, economic and environmental benefits for land users at all levels.

Biodiversity – the variability among living organisms from all sources including, *inter alia*, terrestrial, marine and other aquatic ecosystems, and the ecological complexes of which they are part; this includes diversity within species, between species and of ecosystems.

Biophysical factors – biological, ecological and physical characteristics of a forest landscape that affect restoration programmes.

Conflict – a situation of disagreement between two or more different stakeholders or stakeholder groups.

Coppice management – a forest management system whereby trees are cut back to near ground level every few years to provide straight stems for fuelwood, tools and other purposes.

Degraded forest land – former forest land severely damaged by the excessive harvesting of wood and/or NTFPs, poor management, repeated fire, grazing, or other disturbances or land uses that damage soil and vegetation to a degree that inhibits or severely delays the re-establishment of forest after abandonment.

Degraded primary forest – a primary forest in which the initial cover has been adversely affected by the unsustainable harvesting of wood and/or NTFPs so that its structure, processes, functions and dynamics are altered beyond the short-term resilience of the ecosystem; in other words, the capacity of these forests to fully recover from exploitation in the near to medium term has been compromised.

Direct seeding – the establishment or re-establishment of forest by the direct application of seeds of forest plant species to the soil.

Disturbance – any event that alters the structure, composition or functions of a forest landscape.

Double filter – the principle that the joint objectives of enhanced ecological integrity and human well-being cannot be traded off against each other at a landscape level.

Ecological corridors – strips of habitat that link isolated protected areas. Ecological corridors are not under intensive land use, are distinctly different from surrounding areas, and allow the movement of plant and animal species between protected sites.

Ecological restoration – restoration that aims to closely replicate the structure and floristic composition of the original forest cover and restore the ecological processes and biodiversity to a previous state.

Ecological services (or **environmental services**) – the range of ecological services provided by forests, including hydrological regulation, slope stabilization, soil fertility maintenance, carbon sequestration, and the provision of habitat and gene pools for biodiversity conservation.

Ecological stepping stones – similar to ecological corridors but without their structural continuity, ecological stepping stones are patches of relatively intact habitat between isolated protected areas that allow for species movement, particularly that of mobile species.

Enriched (improved) fallow – a fallow (cultivated land left unseeded for one or more growing seasons) that is managed to accelerate the process of rehabilitation with a view to improving future crop productivity and/or increasing the direct economic benefits of the natural fallow vegetation.

Enrichment planting – the planting of desired tree species in a modified natural forest or secondary forest or woodland with the objective of creating a high forest dominated by desirable (i.e. local and/or high-value) species.

Environmental services – see *ecological services*.

Environmental security – the level of protection of people living in vulnerable locations from the impacts of catastrophic events (including natural disasters, economic shocks or violent conflicts) and the level of access they have to mitigation measures after such events occur.

Estate crop – a crop grown in a plantation system, such as rubber, coffee, tea or oil palm.

Forest degradation – the reduction of the capacity of a forest to produce goods and services. 'Capacity' includes the maintenance of ecosystem structure and functions.

Forest fragmentation – the splitting of continuous forest tracts into a series of smaller patches, due to roads, clearing for agriculture or other human-induced impacts. This process reduces the biodiversity value of the forest.

Forest functionality – the ability of a forest to provide goods and services and maintain ecological processes.

Forest landscape restoration (FLR) – a process that aims to restore ecological integrity and enhance human well-being in deforested or degraded forest landscapes. Rather than aiming to return forests to their previous 'pristine' condition, forest landscape restoration aims to strengthen the resilience and functionality of the forest landscape and keep future forest management options open.

Fuelbreak – strips of land in which flammable material, particularly grasses, has been removed or reduced with the aim of reducing the likelihood of fire spreading from one area to another.

Gap planting – a silvicultural technique, used as part of enrichment planting, whereby seedlings are planted in natural or artificial gaps in the forest cover (see also *enrichment planting* and *line planting*).

Girdling – a method of killing unwanted trees by making continuous incisions through the bark and underlying tissues around the tree stem.

Hydrological processes – ecological processes related to the water cycle, including evaporation, precipitation, water storage, run-off and groundwater flow (see also *ecological services*).

Key informants – individuals selected on the basis of their knowledge, experience or other characteristics to provide information on a particular topic.

Keystone species – a strongly interacting species with a greater influence on ecosystem functioning (including species diversity) than would be predicted based on its abundance. Named after the keystone at the top of an arch structure; if the keystone is removed, the arch collapses.

Land tenure – see *tenure*.

Landscape – a land-area mosaic of interacting ecosystems, land uses and social and economic groupings. It should be noted that a landscape is not necessarily defined by its area; in the context of FLR, the size of the landscape is determined more by the scale of the FLR initiative and the likely or desired geographic extent of its impacts.

Landscape dynamics – changes in the composition of the landscape and changes in the condition of individual components (such as conversion of agricultural land from grazing to crop production).

Landscape mosaic – a patchwork of different components, pieced together to form an overall landscape. The actual composition of the mosaic and the pattern in which the components are distributed will be unique to each landscape.

Liberation thinning – cutting that releases young seedlings, saplings and trees in the canopy C-layer from overhead competition.

Line planting – as with *gap planting*, a silvicultural technique used as part of enrichment planting whereby seedlings are planted along cleared lines (see also *enrichment planting*).

Live fuelbreak – strips of land in which dead plant material and flammable plants are removed along the edge of existing forests and shrublands and the planting of trees at a close spacing to achieve rapid crown closure and early suppression of grasses.

Logical framework – a tool for planning programmes and projects. By leading planners step-by-step through the cause–effect relationships between activities, outputs and goals, it helps link programme inputs and objectives in a clear, logical way and can serve to guide subsequent implementation, monitoring and evaluation activities.

Monitoring and evaluation (M&E) – ongoing and periodic assessments of forest management activities to determine the extent to which objectives are being met, identify any changes in the forest condition and gauge the need for any adjustments in management practices (see also *adaptive management*).

Monoculture – the cultivation of a single tree species in a given area.

Monocyclic system – under a monocyclic system all the marketable volume of timber is harvested in a single felling operation and the production of the subsequent crop relies almost entirely on newly recruited seedlings (see also *polycyclic system*).

Natural regeneration – renewal of trees by self-sown seeds or natural vegetative means.

Ngitili – acacia-miombo woodlands used as a traditional land management system in Tanzania to provide dry season fodder, firewood, and other goods and services.

Non-timber forest products (NTFPs) – all forest products except timber and wood, including products from trees, plants and animals in the forest area.

Passive restoration – a forest restoration strategy that relies largely on protecting the site from the main disturbance or stress factors and allowing natural colonization and successional processes to occur.

Phenological characteristics – the characteristics of tree species that relate to periodic biological phenomena such as leafing and flowering.

Pioneer tree species – heavily light-demanding and short-lived species that can rapidly invade large canopy gaps in disturbed natural forests and colonize open land.

Plantation – a deliberately cultivated area of trees, either with only one species (monoculture) or with a mixture of species (polyculture).

Polycyclic system – under a polycyclic system, commercial trees are repeatedly harvested in a continual series of felling cycles and the production of the subsequent crop relies on the existing stock of seedlings, saplings and poles in the forest (see also *monocyclic system*).

Primary forest – forest which has never been subject to human disturbance or has been so little affected by hunting, gathering and tree-cutting that its natural structure, functions and dynamics have not undergone any changes that exceed the elastic capacity of the ecosystem.

Regrowth forest – see *secondary forest*.

Rehabilitation – a management strategy applied to degraded forest land that aims at restoring the capacity of a forest to produce products and services.

Residual stand – forest that remains after harvesting and extraction.

Resprouts – new growth produced by certain tree species after damage from cutting or fire; such trees may resprout from the trunk, stump or crown.

Riparian (riverine) strips – strips of natural forest retained along streams and rivers, even when the overall land use changes to a non-forest system.

Roundwood – the bole of a tree after felling, usually cut into lengths but otherwise not yet processed into other products such as sawnwood, plywood or veneer.

Scarification – a site preparation technique that mixes the top vegetation with the underlying mineral soil, either by hand or using heavy machinery, to assist regeneration or facilitate planting; scarification reduces competing vegetation, redistributes slash and exposes the mineral soil.

Secondary forest – woody vegetation regrowing on land that was largely cleared of its original forest cover (i.e. carried less than 10 per cent of the original forest cover). Secondary forests commonly develop naturally on land abandoned after shifting cultivation, settled agriculture, pasture or failed tree plantations.

Secondary vegetation – a more general term than *secondary forest*, secondary vegetation can include non-woody regrowth, including shrubland and grassland.

Silviculture – the art and science of producing and tending forests by manipulating their establishment, species composition, structure and dynamics to fulfil given management objectives.

Sliding strip planting – a variation of *line planting*, based on successive expansion of alternate sides of a planted strip.

Social capital – a concept based on the idea that people's social networks are a valuable asset.

Stakeholder – any individual or group directly or indirectly affected by, or interested in, a given resource. In the context of FLR we define a stakeholder as an individual, group of people or organization that can directly or indirectly affect an FLR initiative or be directly or indirectly affected by it.

Stakeholder approach – an approach (in this context to forest management) that seeks to identify and understand the needs and concerns of the different *stakeholder* groups associated with the forest resource in question and to work with these groups to plan, implement and monitor an appropriate restoration programme.

Succession – progressive change in species composition and forest structure caused by natural processes over time.

Swidden agriculture (shifting agriculture) – a traditional farming system that involves clearing small areas of forest to grow crops and/or raise livestock, leaving a fallow period of varying length to enable forest regeneration; swidden agriculture becomes unsustainable when viable fallow periods are not maintained.

Tenure – agreements held by individuals or groups, and recognized by legal statutes and/or customary practice, regarding the rights and duties of ownership, holding, access and/or usage of a particular land unit or the associated resources (such as individual trees, plant species, water or minerals) therein.

Trade-off – a situation where a balance needs to be reached when choosing between two desirable but incompatible objectives or outcomes.

Understorey – the plants growing beneath the forest canopy, often including grasses, shrubs, vines, ferns, etc. The understorey is usually the most diverse layer of the forest and is important for biodiversity, tree survival and soil fertility.

Wildlings – naturally sown seedlings collected for planting elsewhere.

Index